BEWARE OF THE FEAST
The History of Robt. Jowitt & Sons

Fig. 1 Robert Jowitt (1784–1862)

BEWARE OF THE FEAST
The History of Robt. Jowitt & Sons

Peter Danckwerts

WB

RICHMOND: THE WORD BUSINESS
2011

First published in the UK in 2011 by
The Word Business Ltd
50 Albert Road
Richmond
Surrey TW10 6DP

ISBN 978-0-9570010-0-8

Designed & typeset in the UK by
The Word Business Ltd
Printed & bound in the UK, US and Australia by
Lightning Source

Contents

CONTENTS

List of Illustrations

Introduction

Robt. Jowitt & Sons Ltd celebrated its bicentenary in 1975. This was both too early and too late. The Jowitt family had been involved in the wool trade in the early eighteenth century and probably much earlier but it did not become Robt. Jowitt & Son until 1833 and Robt. Jowitt & Sons plural until about 1842. It only became a limited company in 1919. Nonetheless, the year 1775 is significant in the history of the firm, as the oldest surviving business records for the family business date from then.

This book was commissioned by Mr F. T. B. Jowitt, the last Chairman of Robt. Jowitt & Sons Ltd. It is not intended to be, and could not be, a narrow company history. The Jowitts' commercial activities took place against a background of religious, social and personal events from which they cannot be divorced. For the period up to the 1830s (and in the case of Robert up to the 1860s), membership of the Society of Friends dictated how the Jowitts carried on their business, whom they married and, to some extent, with whom they traded. Both their consciences and commercial interests dictated their attitudes and actions with regard to politics, religion and social reform. The family was at the forefront of the abolitionist and temperance movements. They were vocal in their support for the Reform Bill and equally vocal in their criticism of the opium trade. Members of the Jowitt family were not only deeply shaken by the Beaconite schism in the British Quakers of the 1830s, they were amongst the most active participants on both sides of the debate, with John junior suggesting changes to the second edition of the *Beacon* to his uncle Isaac Crewdson and his father writing a rebuttal of Crewdson's remarks on baptism.

Even if a purely commercial history had been desirable, the material available would have made it difficult. Although there are many ledgers, they are not always very informative. The business correspondence files, though occasionally interesting, are patchy and hard to read; worse than that, they are often unsigned and unaddressed, and sometimes undated. There is no clear divide between business and personal correspondence, or even between business and personal finances. Inevitably, much of the material has been lost, but what romance would there be in History if every document survived? The minute books of Quaker Meetings and many newspaper reports, and a few diaries which have survived, have helped to flesh out both the company history and the underlying human drama.

The diaries of Dearman Birchall and his wife Emily (*née* Jowitt) published in *Diary of a Victorian Squire* and *Wedding Tour* provided interesting insights into Victorian life, and I hope that readers will not find the extended extracts given in this book from the diaries of Emily's father, John Jowitt junior, out of place. Although there is no suggestion of impropriety, it is remarkable that in 1835 John and his future wife, Deborah Benson, effectively took their long Continental honeymoon *before* their wedding – before they were even officially engaged.

There have been Jowitts in Yorkshire for many centuries. In early years, the name was spelt Jouet, Jowet, Jowett, Jowitt or even Joiet. By the early eighteenth century, the family we are concerned with had apparently standardised the spelling to Jowitt, but it is more than likely that some of those who spelt their name Jowett or Jowet are related to them. However, unless there is clear evidence, it is very unwise to assume that any Jowitt (however it is spelt) is related to the family which founded Robt. Jowitt & Sons. For instance, there was a Richard Jowet, a clothier of Holbeck with a wife called Sarah, who died in 1756 and who had a brother Joseph, also of Holbeck, a maltster. One might assume that they are John Jowitt I's brothers Richard (b. 1725) and Joseph (b. 1737), both born in Holbeck, which is and was, after all, a very small place. However, Richard Jowitt who was John Jowitt I's brother lived until 1766 and was married to a Paulina Brigg.

A NOTE ON DATES, SPELLINGS AND TRANSCRIPTIONS

Dates

Until 1752, the year in which Britain and her colonies rather belatedly changed to the Gregorian calendar, the official year in Britain began on 25 March – that is, 31 December 1683 was followed immediately by 1 January 1683, and 24 March 1683 was followed by 25 March 1684. An English document dated 1 January 1683 is therefore usually 1684 according to modern dating. I have adopted the common practice of double-dating thus: '1 January 1683/4'. Wednesday 2 September 1752 (which was 13 September 1752 in most of Europe) was followed by Thursday 14 September 1752 in order to bring Britain into line with the Continent.

Quakers, including the Jowitts, long avoided the use of heathen names for the months, referring instead to 1st month (or 1st mo.), 2nd month, etc. However, British Quakers started their year in March like everyone else, so 2nd month is April, not February. To confuse matters further, Quakers tended to treat the whole of March as if it belonged to the new year, so 24 1 mo. 1684 *probably* refers to 24 March 1683/4, but it may refer to 24 March 1684/5. Equally offensive to the Quakers were the heathen names of the days of the week, so Sunday is usually referred to as 1st day, etc. The Jowitts clung on to these traditions, as well as thee-thouing their correspondents, well into the nineteenth century and, in Robert's case, until the end of his life in 1862. Even those, like John junior, who had left the Society of Friends, sometimes retained Quaker dating styles.

The main problem with dating in this book has been that it is hard to know whether some secondary sources, such as Besse's *Sufferings,* published after the calendar change, have correctly converted the old dating to the new. Dates of births, marriages and deaths quoted in genealogical works are frequently incorrect.

Names

The name Jowitt was often spelt Jowett in the seventeenth and early eighteenth century but I have standardised the spelling to Jowitt, except in direct quotations. Many other personal and surnames vary from one document to another and I have tended to stick with the most frequent usage.

The personal names adopted by the Jowitt family can cause some confusion. Not only were there many Roberts, two Richards and three Johns in direct descent, there were cousins and unrelated or distantly-related Jowitts with the same names. At the time, some of the confusion was avoided by calling people by the middle names (although middle names were generally eschewed by early Quakers as a wasteful and frivolous affectation). Thus, Robert Benson Jowitt was known to his family as Benson and Robert Crewdson Jowitt as Crewdson. Unfortunately, this creates, for the modern reader, as many problems as it solves. Does Crewdson refer to Robert Crewdson Jowitt or to his uncle Isaac Crewdson? Does Benson refer to Robert Benson Jowitt or to *his* uncle William Thomas Benson? I have, with some reluctance, since it does not reflect contemporary usage, numbered some of the family. Therefore, John

Jowitt I (or simply John I) refers to John Jowitt (1721–1783), although, for various reasons, the third John in direct descent is referred to as John junior, not John III, and his cousin as John senior.

With equal reluctance, I have referred to others, when not giving their full names, by their initials – for instance, RCJ for Robert Crewdson Jowitt. The reader must remember, however, that when a member of the family is quoted as referring to 'Benson' or 'Crewdson', they mean Robert Benson Jowittt or Robert Crewdson Jowitt. Some of the names which might lead to confusion are listed below:

> John Jowitt I (son of Richard II) 1721–1783
> John Jowitt II (son of John I) 1750–1814
> John Jowitt senior (son of Joseph Jowitt senior) 1790–1860
> John Jowitt junior (son of Robert Jowitt) 1811–1888
> Joesph Jowitt senior (brother of John Jowitt II) 1757–1803
> Richard Jowitt I 1661–1696
> Richard Jowitt II (son of Richard I) 1694–1741
> Richard Jowitt III (son of John I) 1748–1765
> Robert Jowitt (son of John II) 1784–1862
> Robert Jowitt II (son of RBJ) 1870–1945
> Robert Benson Jowitt *or* RBJ *or simply* Benson
>   (son of John junior) 1839–1914
> Robert Benson Jowitt II (son of Frederick McC. Jowitt)
>   1901–1966
> Robert Crewdson Jowitt *or* RCJ *or simply* Crewdson
>   (son of Robert) 1821–1847
> William Thomas Benson (son of Robert Benson) 1824–1885

Transcriptions

In transcribing extracts from old documents, the spelling has not been modernised, but abbreviations and contractions have been expanded where they might cause the modern reader difficulties – for instance, the use of y^t for *that* in Quaker minute books.

ACKNOWLEDGEMENTS

I am very grateful to the Jowitt family for their invaluable help in writing this book, especially Mr F. T. B. ('Tommy') Jowitt and his wife Juliet for their generous hospitality and unstinting help. Tommy has been a most valuable source for information on the undocumented history of the company and the lore of the Jowitts. Wholehearted thanks are also due to the Reverend David Jowitt and the Reverend Andy Jowitt (and his wife Jane) for giving me valuable background information on their side of the family, especially J. H. Jowitt and his son F. R. B. Jowitt. I am also extremely grateful to Mr R. E. Jowitt who gave me some fascinating background on his branch of the family, including the Simpsons and the Benedicts, and who allowed me to photograph his many fascinating pictures and documents.

The most industrious historian of the Jowitt family has been Jason Jowitt in Australia and I am extremely grateful to him for allowing me to plunder his researches. He is descended from Joseph Jowitt senior (1757–1803) whose branch of the family ceased to be involved with what was to become Robt. Jowitt & Sons Ltd in 1802. As a result, it has been difficult to fit much of the information on his descendants into the main body of the text. However, since they include both the former Lord Chancellor, Lord Jowitt, and the eccentric millionaire Robert Arthington, they could not be omitted entirely and I have included some of them in an appendix. All the Jowitts I have met have been charming and amusing, making the work on the book a most enjoyable experience.

I also owe a great debt to Mr Peter J. M. Bell, the former Managing Director of Robt. Jowitt & Sons Ltd, for providing me with useful background information on the the Bradford wool industry in general as well as Jowitts itself. He has written two mongraphs without which it would have been difficult or impossible to complete my own book: *Robert Jowitt & Sons, the Limited Company 1919 to 1957* and *A History of the Woolcombers' Mutual Association Limited 1933 to 1994*. The former has illuminated for me obscure references in the company's minute books and the latter has likewise elucidated the rather murky history of the WMA. While I have barely touched on the WMA in this book, it is a subject of great significance in the history of the British woolcombing industry and I recommend Mr Bell's book to anyone working in that field.

I have drawn on another history of the company, that compiled by Mr R. E. Jowitt's mother Dorothy. This has been particularly useful since she referred to a number of sources which seem to have disappeared.

No history of Robt. Jowitt & Sons would have been possible without the help of the Brotherton Library of the University of Leeds. The

Library's Special Collections contain not only many of the company's archives, deposited on permanent loan in 1949[1] but also a wealth of other material, including Quaker minute books which shed light on the early history of the Jowitt family. Anyone who has had the privilege to work in the Brotherton will know what a very special library it is. The subterranean levels are admittedly a little cramped and claustrophobic but the quality of the collections and the helpfulness of the staff more than make up for this. I would particularly like to thank Mr Chris Sheppard and his staff in Special Collections (which, by the way, are housed in rooms which are neither cramped nor claustrophobic) who are unfailingly helpful and knowledgeable and who make studying there a particularly productive and enjoyable experience.

Leeds, with so much of its rich and fascinating past cruelly obliterated by bulldozers in the postwar years, deserves a vibrant local history society and fortunately it has one in the Thoresby Society. I would like to thank its staff for their generous help in hunting down background reading for me. The Leeds City Library's local history collection is also very good and its staff extremely helpful. Further to the north, the Cumbria Archives at Kendal have proved themselves both helpful and very efficient. I must also thank the Open University, without whose online databases this book would have been impossible.

My thanks also to Tom Sykes and the Reid family – Sharon, Samantha and David – for their valuable information of the Robt. Jowitt & Sons South African subsidiary.

I must thank my sister Ingrid and my old friend Ben Andrews for reading the minutes and various drafts. Last but not least I must thank my wife Elizabeth for holding things together as I went to pieces.

PICTURE CREDITS

All pictures courtesy of F. T. B. Jowitt except: 2 Borthwick Institute, University of York; 3, 7, 10 University of Leeds Library; 6, 13, 21, 22, 29, 30, 31 Robert Ernald Jowitt; 8 Peter Danckwerts; 9 Dix Noonan Webb; 14, 17, 18, 28 Rev. David Jowitt; 15, 16, 19, 24, 34 Rev. Andrew Jowitt; 20 Leeds Library & Information Services; 27 Topley Studio/Library & Archives Canada; 39 Paramount Pictures; 43 Sharon Reid; 46 Juliet Jowitt.

The Waste Land

From Hell, Hull and Halifax, Good Lord preserve us.

[ancient Yorkshire saying]

... there are at least 600,000 waste acres in the single county of Northumberland. In those of Cumberland and Westmoreland, there are as many more. In the north and part of the west riding of Yorkshire, and the contiguous ones of Lancashire; and in the west parts of Durham are yet greater tracts.

Young, *Observations*, 1773[2]

LEEDS

To Arthur Young and many other observers, much of the north of England, including the West Riding, was waste land in need of enclosure and 'improvement'. What they failed to notice, or perhaps *chose* not to notice, was the dynamic cloth industry which had grown up in the apparently inhospitable and unproductive moorlands around Leeds, Halifax, Huddersfield and Bradford. Here, and in many other parts of England, notably the West Country, wool had become the great generator of wealth. Even that most perceptive of observers, Daniel Defoe, described it as 'this otherwise frightful Country,'[3] the 'otherwise' referring to its natural suitability for the wool trade. Approaching Leeds from Halifax in the early seventeenth century, he noted that 'every way to the right hand and the left, the country appears busy, diligent, and even in a hurry of work.' He was much impressed by the 'noble scene of Industry and Application' centred upon Leeds.[4] At this time, and until the end of the eighteenth century, cloth-making was a fairly small-scale industry. Of the manufacturers it was said that they were

generally men of small capitals, and often annex a small farm to their other business. Great numbers of the rest have a field or two, to support a horse and a cow; and [were], for the most part, blessed with the comforts, without the superfluities of life.[5]

It should be noted, however, that farming was considered ancillary to their main business of producing wool cloth, and even in the early eighteenth century wool manufacture was something slightly more than

7

a cottage industry as is clear from Defoe's description of the trade outside Halifax:

> ... if we knocked at the door of any of the Master Manufacturers, we presently saw a house full of lusty Fellows, some at the Dye-fat, some dressing the Cloths, some in the Loom, some one thing, some another, all hard at work, and full employed upon the Manufacture, and all seeming to have sufficient Business.[6]

Whereas the cloth trade centred on Halifax consisted of scattered dwellings, that around Leeds was clustered in villages which tended to specialise in the manufacture of either white or coloured cloth. The main villages in which white cloth was produced were Alverthorpe, Ossett, Kirkheaton, Dewsbury, Batley, Birstal, Hopton, Mirfield, Cleckheaton, Littletown (near Liversedge), Bowling and Shipley. 'Mixture' (woven from dyed yarn) was produced in some of the villages belonging to the town of Leeds and in Morley, Gildersome, Adwalton, Drighlington, Pudsey, Farsley, Calverley, Eccleshill, Idle, Baildon, Yeadon, Guiseley, Rawdon, Horsforth, Batley, Dewsbury, Ossett, Horbury, and Kirkburton.[7]

The cloth markets were originally held on the stone bridge over the river Aire but rapid expansion and the traffic congestion caused by the market forced a move north to Briggate, then the town's principal street, in 1684. In response to a covered cloth hall erected in Wakefield in 1710, the first White Cloth Hall was built in Kirkgate in 1711, followed by a Coloured Cloth Hall near Mill Hill. Continued expansion required the building of a larger White Cloth Hall south of the river in 1756 and a larger still on the south side of Call Lane in 1776. It was ninety-nine yards long and seventy wide, divided into five 'streets', each containing two rows of stands, the total number of stands being 1210.[8] Above the Cloth Hall was an elegant assembly room for balls and other public functions. The third White Cloth Hall was demolished in 1865 to make way for a railway viaduct and was replaced by a fourth in King Street.

In the eighteenth century, markets were held twice a week: on Tuesdays for white cloth and on both Tuesdays and Saturdays for coloured cloth. Generally the cloth sold here was in a rough state, straight from the fulling mills, and would be finished off by dressers and dyers employed by the merchants who purchased it.

Defoe observed the impressive organisation of the cloth market at Briggate:

Early in the Morning, there are Tressels placed in two Rows in the Street; sometimes two Rows on a side, but always one Row at least; then there are Boards laid cross those Tressels, so that the Boards lie like long Counters on either Side, from one end of the Street to the other.

The Clothiers come early in the Morning with their Cloth; and as few Clothiers bring more than one Piece, the Market being so frequent, they go into the Inns and Publick Houses with it, and there set it down.

At seven a Clock in the Morning, the Clothiers being supposed to be all come by that time, even in the Winter, but the Hour is varied as the Seasons advance (in Summer earlier, in the Depth of Winter a little later) I take it, at a Medium, and as it was when I was there, at six or seven, I say, the Market Bell rings; it would surprize a Stranger to see how few Minutes, without hurry or noise, and not in the least disorder, the whole Market is fill'd; all the Boards upon the Tressels are covered with Cloth, close to one another as the Pieces can lie long ways by one another, and behind every Piece of Cloth, the Clothier standing to sell it...[9]

The market bell would ring again at half past eight and all the buyers would leave immediately. According to Defoe almost all the cloth would be sold, at a total value of ten or twenty thousand pounds per market day, sometimes much more.

As the industrial revolution took hold, Leeds also profited from a growing engineering business, including the manufacture of steam engines, and sales of locally-mined coal.

Neither the wool industry, nor the newer industrial concerns could flourish without an efficient transport network. The cloth trade was not provincial, relying on the outside world both for its raw material and for its market. By the early eighteenth century, raw wool was brought into Leeds from all over England, from Germany and Spain, and the cloth sold as far afield as Hamburg, Holland, Frankfurt, Leipzig, Vienna, Augsburg and even (via London) St Petersburg, Sweden, Pomerania, Riga, New York, Virginia and New England.[10] Leeds, land-locked and set in a rugged landscape, had started at a serious disadvantage. In the seventeenth century, the main road from Chester passed through Manchester and Elland, near Halifax, to Leeds and on to York. It was hilly and badly maintained, so that packhorses rather than wagons were required for part of its length, meaning that transport was both slow and expensive. As Leeds and the other manufacturing towns were not self-sufficient in food, butter, beef, mutton and grain also had to be brought in by road. In the case of beef, black cattle were walked in from the North Riding in September and October and were then slaughtered, salted and smoked to preserve the meat for the rest of the year.[11]

Since before the beginning of the seventeenth century, Leeds merchants had argued for the rivers Aire and Calder to be made navigable, but this had been opposed by vested interests, including the burghers of York, who had been granted powers over the Ouse and its tributaries in a charter of 1462. In particular, the clothiers of York stood to lose trade if the scheme went through. For a long time, also, the cost of the enterprise seemed prohibitive. Ironically, it was a new tax of four shillings per chaldron[12] imposed on sea-borne coal in 1695 which ensured the scheme's viability, for it gave the transport of coal by river from Leeds to York an unfair advantage over coal sent by sea from Newcastle. By 1704, the Aire and Calder Navigation Company had made the river Aire navigable as far as Leeds by the construction of locks and lock cuts. Within two years, the company had carried out similar work on the Calder. Ships that left with cloth and coal would return with butter, cheese, salt, sugar, tobacco, fruit, spices, hops, oil, wine, brandy, spirits, lead, iron and other heavy or bulky goods.[13] Water transport became even more important to Leeds with the construction of the Leeds and Liverpool Canal in the 1770s.

Despite, or perhaps because of, the increased competition from the waterways, the road system was greatly improved during the eighteenth century, especially after the Yorkshire Turnpike Act of 1734.[14] The journey by packhorses between Leeds and London took seven days in 1729. By 1813, wagons could make the journey in five days, and by 1838 they were making it in three days.[15]

The railways came early to Leeds with the Middleton Railway being constructed in 1758. Ironically, the trains, used for hauling coal, were originally pulled by horses. However, steam locomotives produced by the Leeds firm of Fenton, Murray and Wood were introduced in 1812.

Despite the improved transport links and burgeoning industry, Leeds in the late eighteenth century was not very different from the 'pretty market-town, reasonably well builded; as large as Bradford, but not so quick as it,' described by Leland in the sixteenth century.[16] Perhaps the most obvious changes were the new public buildings such as the cloth halls and the magnificent General Infirmary opened in 1771 and extended in 1785, a far more attractive and less fussy-looking structure than the Victorian hospital which replaced it. It was described as

> a very handsome and spacious edifice, situated in a large and pleasant Square, at the West end of town. It is surrounded by a large court and garden, and is accommodated with every requisite out-building... This Institution is, as its name imports, a General Infirmary, to which every proper object, whatever his place of residence, has an equal claim, provided he is recommended by a Subscriber; but in case of accidents or cases not admitting of delay, this Recommendation is dispensed with. The only diseases excluded from the benefit of this Institution are the venereal disease and infectious diseases, for the relief of the latter the House of Recovery has been erected.[17]

The social reformer John Howard, visiting the Infirmary in 1788, declared:

> This is one of the best Hospitals in the kingdom. In the wards, which are fifteen feet eight inches high, there is a great attention to cleanliness, and six circular apertures, or ventilators open into a passage five feet and a half wide. Many here are cured of compound fractures, who would lose their limbs in the unventilated and offensive wards of some hospitals.[18]

Although warehouses, public buildings and handsome new merchants' houses had been erected, much of the medieval architecture remained, and the population in 1775 was little more than seventeen thousand.[19]

Yorkshire had long been a stronghold of non-conformists and Leeds contained many chapels for dissenters, including the Mill Hill Chapel, erected by the Unitarians in 1672, which was presided over for some years by Joseph Priestley, who isolated oxygen (or 'dephlogisticated air' as he called it) and wrote *The History and Present State of Electricity*. A later writer recorded that the chapel was 'incrusted in grey plaster, and... shaded by large trees, and well accords with the solemnities of public worship.'[20]

The Albion Chapel, erected as an Anglican house of worship, was bought by the Presbyterians in 1802. The Call Lane Chapel, built in 1691, had a congregation of Independents 'without any tincture of calvinism,'[21] while the White Chapel, which was 'situated in a very unpleasant and confined part,'[22] housed the Calvinist Independents. Both the handsome Salem Chapel and the Bethel Chapel in St George's Street also contained Independents. There were also two Methodist chapels and an Ebenezer Methodist chapel, a Roman Catholic church and the Inghamite chapel in Duke Street which had fallen into disuse by 1806.

The Society of Friends was founded during the 1640s and in 1699 the first purpose-built Quaker meeting house was erected in Leeds at the top of Camp Lane, Holbeck. However, there seems to have been a burial ground there at least thirty years before that.[23] The original meeting house fell into disrepair and a new one was built of stone in 1788.

Ever more exotic sects appeared, including the 'obstreperous sisterhood'[24] of Female Methodist Revivalists who erected their chapel in Regent Street, Leylands, in 1825. Perhaps the most exotic of all were a species of Primitive Methodists known as the Jumping Ranters who worshipped at the Rehoboth Chapel and

> who perform their gymnastic devotions most assiduously, congregating three times every Sunday, besides certain evenings during the week, when the constables are often under arms, in order to preserve some degree of subordination, the uproar being full as great without the chapel as within, the zeal of their holy dancings bringing together a most disorderly mob.[25]

The existence of many chapels and meeting houses for dissenters in Leeds should not be taken to indicate that they were welcomed or even tolerated by the authorities. Various acts of parliament made life difficult for dissenters, particularly the Quakers. The Quaker Act of 1662 prohibited more than four Quakers from meeting in worship and made it a crime to hold an oath unlawful or even to persuade others not to take an oath.[26] The Conventicle Acts of 1664 and 1670 reaffirmed the illegality of Quaker meetings and other unauthorised religious gatherings. Charles II had attempted to introduce religious toleration by issuing a Declaration of Indulgence in 1672 and James II issued a similar one in 1687. In 1689, under William and Mary, the Act of Toleration (An Act for Exempting their Majestyes Protestant Subjects dissenting from the Church of England from the Penalties of certaine Lawes, 1 Will. & Mar. c. 18) was passed, granting freedom of worship to dissenting protestants.

From then on, Quakers gradually achieved a more comfortable position in British society.

KENDAL

About fifty-six miles north-west of Leeds, as the crow flies (but over seventy miles by road, even today), lies the once picturesque town of Kendal, in what was until 1974 the county of Westmorland. Despite the distance, the Jowitts of Leeds were to establish close links with Kendal.

Although generally described as rich,[27] Kendal had a 'large and commodious workhouse for the poor'[28] and one traveller reported that, although the air was good, the children of Kendal looked

> very sickly, but in the neighbourhood they are a rosy race: perhaps they suffer from the nap of the woollen manufactory, which is constantly flying about, clogging their infant lungs.[29]

Until it was popularised by the romantic poets and painters, many outsiders regarded Westmorland as a fearsome place. The rugged beauty of its landscape made Kendal relatively inaccessible. The transport links were improved first by a turnpike road and then in 1819 by the completion of the Lancaster Canal. Mrs Radcliffe, the celebrated author of Gothic horrors, described vividly both the wild scenery and difficult road journey in 1794:

> Leaving Kirby-Lonsdale by the Kendal road, we mounted a steep hill, and, looking back from its summit upon the whole vale of Lonsdale, perceived ourselves to be in the mid-way between beauty and desolation, so enchanting was the retrospect and so wild and dreary the prospect. From the neighbourhood of Caton to Kirby the ride was superior, for elegant beauty, to any we had passed; this from Kirby to Kendal is of a character distinctly opposite. After losing sight of the vale, the road lies, for nearly the whole distance, over moors and perpetually succeeding hills, thinly covered with dark purple heath flowers, of which the most distant seemed black. The dreariness of the scene was increased by a heavy rain and by the slowness of our progress, jostling amongst coal carts, for ten miles of rugged ground. The views over the Westmoreland mountains were, however, not entirely obscured; their vast ridges were visible in the horizon to the north and west, line over line, frequently in five or six ranges. Sometimes the interfering mountains opened to others beyond, that fell in deep and

abrupt precipices, their profiles drawing towards a point below and seeming to sink in a bottomless abyss.

On our way over these wilds, parts of which are called Endmoor and Cowbrows, we overtook only long trains of coal carts, and, after ten miles of bleak mountain road, began to desire a temporary home, somewhat sooner than we perceived Kendal, white-smoking in the dark vale. As we approached, the outlines of its ruinous castle were just distinguishable through the gloom, scattered in masses over the top of a small round hill, on the right. At the entrance of the town, the river Kent dashed in foam down a weir; beyond it, on a green slope, the gothic tower of the church was half hid by a cluster of dark trees; gray fells glimmered in the distance.[30]

Like Leeds, Kendal had a large Quaker population, and one of its main industries was the the production of coarse woollen cloth from the local Westmorland wool, known misleadingly as 'Kendal cottons' (even the term 'Manchester cottons' originally referred to woollen cloth). Kendal also produced other woollen and worsted cloths from wool brought in from Lancashire and Yorkshire – druggets and serges – as well as hats, stockings, linen cloth, linsey-woolsey (a coarse linen/wool cloth). The last Kendal guild procession in 1759 included:

Woolcombers, 100; Taylors, 150; Shearman-dyers, 80; Weavers, 300; Shoe-makers, 100; Ironmongers and Mettlemen, 80; Tanners, 60; Builders, 100; Glovers and Skinners, 70; Mercers, &c., not numbered. Strangers were allowed to join in if they requested, and were provided with sashes and cockades, 'provided they had given a fortnight's notice.'[31]

Even gunpowder was produced in Kendal, not an obvious commodity for a pacifist Quaker population, although it must be remembered that it had applications in quarrying and game hunting as well as war, and the Quakers, while numerous, were not in the majority. It was not perhaps as dangerous as that most insidious product, Kendal Black Drop, a painkiller containing opium, vinegar and spices. It was this drug, said to be four times as strong as standard laudanum, which Samuel Taylor Coleridge took for rheumatism and caused his precipitous decline into drug addiction. He wrote,

I had always a fondness (a common case, but most mischievous turn with reading men who are at all dyspeptic) for dabbling in medical writings;

and in one of these reviews I met a case, which I fancied very like my own, in which a cure had been effected by the Kendal Black Drop. In an evil hour I procured it: – it worked miracles – the swellings disappeared, the pains vanished; I was all alive, and all around me being as ignorant as myself, nothing could exceed my triumph. I talked of nothing else, prescribed the newly discovered panacea for all complaints, and carried a bottle about with me, not to lose any opportunity of administering 'instant relief and speedy cure' to all complainers, stranger or friend, gentle or simple... Alas! it is with a bitter smile, a laugh of gall and bitterness, that I recall this period of unsuspecting delusion, and how I first became aware of the Maelstrom, the fatal whirlpool, to which I was drawing just when the current was already beyond my strength to stem.[32]

Devised by a Quaker, Dr Edward Toustall,[33] this 'flattering poison,' as Coleridge called it, is known to have been used by the Jowitts of Leeds, even after the publication of Coleridge's doleful words. Robert Jowitt, who gave his name to the wool company wrote:

Dear Susan[nah] has been confined to bed for about a fortnight, with very severe paroxysms of pain in the lower part of the body, for which she has frequently had to take a dose of 30 drops black-drop, and her head has often also been very bad.[34]

BRADFORD

Bradford is only about seven miles west of Leeds. At the beginning of the nineteenth century it was little more than a market town, long associated with the wool textile industry but not on a very large scale. There were Jowitts in Bradford, including a John Jowitt of Bowling, a Quaker imprisoned in York Castle in 1661 for not swearing the Oath of Allegiance.[35] It is entirely possible that he was related to the Jowitts with whom we are concerned, but there is no evidence of this. *Our* Jowitts, while travelling widely across the country on business, seem to have had very little to do with Bradford until well after its rise as Worstedopolis, the centre of the English worsted trade.

The Jowitts of Leeds

O Ye People of *Leeds!* who have persecuted the Lords Servants, and imprisoned the Innocent without a cause; For your sakes am I made to mourn in secret, and my Spirit is troubled within me, when I consider the Calamity that is coming upon you, and which will certainly overtake you, except you speedily Repent: A Day of Distress and great Lamentation! A Day of Darkness and Gloominess! A Day of Misery and of the Shadow of Death, is coming upon you...

Isabel Wails, 1685[36]

I cannot wholly omit my concern for some poor deluded Quakers, who were hurried down this street to York castle, in greater numbers than was ever known in these parts. The Lord open the eyes of the one party, and tender the hearts of the other!

Ralph Thoresby, 1724[37]

The authorities suspected Quakers of many things, including sedition, and not entirely without cause. Along with many non-conformists, they refused to pay tithes or take the Oath of Allegiance and after the Restoration a number had taken part in uprisings against the Crown, such as the Farnley Wood Plot of 1663 (see p. 23).

On 27 February 1682/3, a Leeds Meeting of the Society of Friends was disrupted by two aldermen, Joshua Balmer and Martin Headley; twenty people were arrested and imprisoned, first in Leeds and then at York Castle. One of those arrested was Richard Jowitt (or Jowett as it was usually spelt at the time), a yeoman of Beeston, a substantial town within the parish of Leeds then known for the production of bobbin lace and straw hats.[38] Elsewhere he is described as a miller at Millshaw, near Beeston, and the mill he occupied was undoubtedly the same one mentioned by Ralph Thoresby in his history of Leeds in 1715.[39] It is clear from an inventory made at the time of his death[40] that Richard was a corn miller.

At the Quarter Sessions held at Leeds on 26 January 1682/3, Richard Jowitt and the others were indicted for holding an assembly of illicit and riotous assembly, for which offence they were later convicted and imprisoned. On 30 May 1684, Richard Jowitt, William Cowel, John Sikes (or Sykes), Bryan Sheffield, Richard Roe, Benjamin Elleston, James Pearson, Joseph Liversedge, Joseph Lupton, Isabel Wails (or Wailes),

THE JOWITTS OF LEEDS

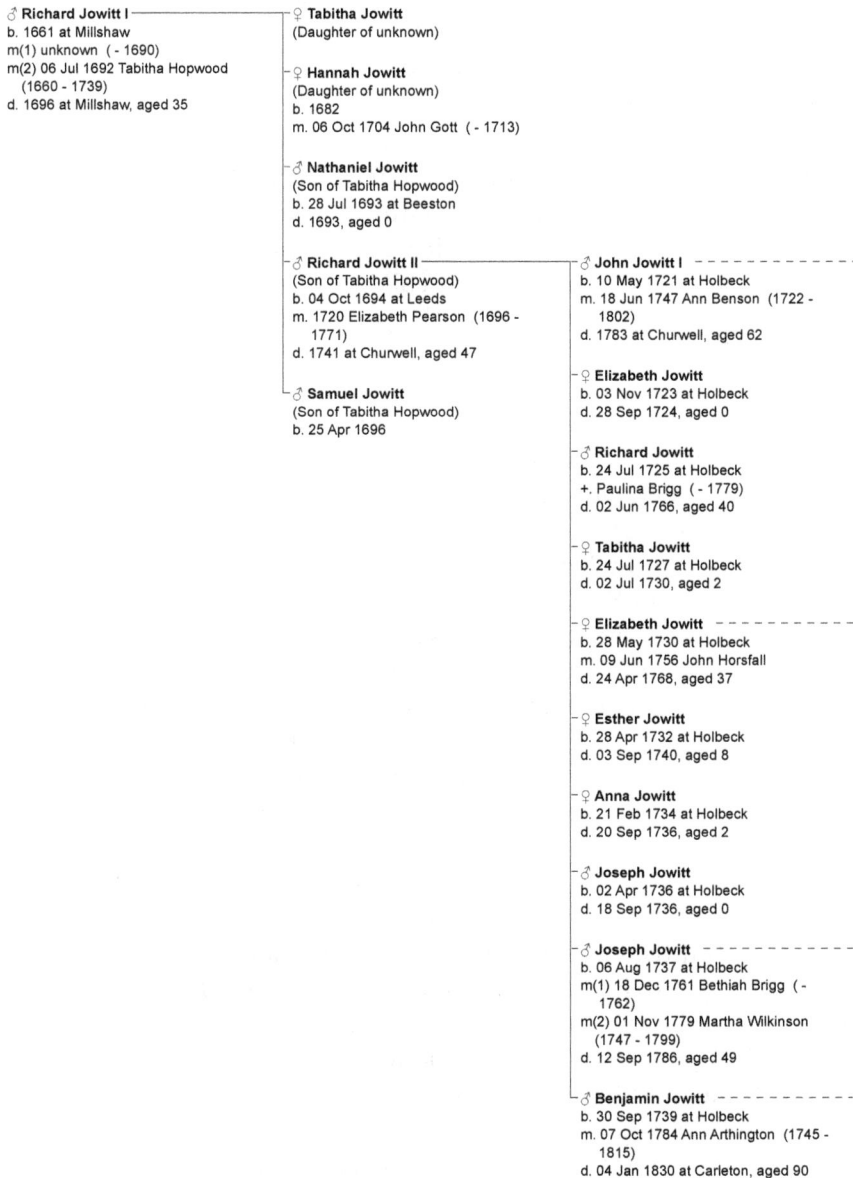

♂ **Richard Jowitt I**
b. 1661 at Millshaw
m(1) unknown (- 1690)
m(2) 06 Jul 1692 Tabitha Hopwood
    (1660 - 1739)
d. 1696 at Millshaw, aged 35

♀ **Tabitha Jowitt**
(Daughter of unknown)

♀ **Hannah Jowitt**
(Daughter of unknown)
b. 1682
m. 06 Oct 1704 John Gott (- 1713)

♂ **Nathaniel Jowitt**
(Son of Tabitha Hopwood)
b. 28 Jul 1693 at Beeston
d. 1693, aged 0

♂ **Richard Jowitt II**
(Son of Tabitha Hopwood)
b. 04 Oct 1694 at Leeds
m. 1720 Elizabeth Pearson (1696 -
    1771)
d. 1741 at Churwell, aged 47

♂ **Samuel Jowitt**
(Son of Tabitha Hopwood)
b. 25 Apr 1696

♂ **John Jowitt I**
b. 10 May 1721 at Holbeck
m. 18 Jun 1747 Ann Benson (1722 -
    1802)
d. 1783 at Churwell, aged 62

♀ **Elizabeth Jowitt**
b. 03 Nov 1723 at Holbeck
d. 28 Sep 1724, aged 0

♂ **Richard Jowitt**
b. 24 Jul 1725 at Holbeck
+. Paulina Brigg (- 1779)
d. 02 Jun 1766, aged 40

♀ **Tabitha Jowitt**
b. 24 Jul 1727 at Holbeck
d. 02 Jul 1730, aged 2

♀ **Elizabeth Jowitt**
b. 28 May 1730 at Holbeck
m. 09 Jun 1756 John Horsfall
d. 24 Apr 1768, aged 37

♀ **Esther Jowitt**
b. 28 Apr 1732 at Holbeck
d. 03 Sep 1740, aged 8

♀ **Anna Jowitt**
b. 21 Feb 1734 at Holbeck
d. 20 Sep 1736, aged 2

♂ **Joseph Jowitt**
b. 02 Apr 1736 at Holbeck
d. 18 Sep 1736, aged 0

♂ **Joseph Jowitt**
b. 06 Aug 1737 at Holbeck
m(1) 18 Dec 1761 Bethiah Brigg (-
    1762)
m(2) 01 Nov 1779 Martha Wilkinson
    (1747 - 1799)
d. 12 Sep 1786, aged 49

♂ **Benjamin Jowitt**
b. 30 Sep 1739 at Holbeck
m. 07 Oct 1784 Ann Arthington (1745 -
    1815)
d. 04 Jan 1830 at Carleton, aged 90

Anne Cooper and Hannah Thackeray appeared before the Quarter Sessions in Leeds and were released after eleven weeks in prison.[41] One of those freed, Isabel Wails, must have been imprisoned again, for on 17 October 1684 she wrote *A Warning to the Inhabitants of Leeds,* quoted above, from York Castle.

Richard's first wife (whose name we do not know) died in 1690, leaving him two daughters, Hannah, born in 1682, and Tabitha (dates uncertain).[42] On 6 July 1692 he married Tabitha Hopwood from Wortley, just West of Leeds. It was normal practice among the Friends to require a widower to make a settlement on any children of his first marriage before remarrying.[43] Perhaps as a result of the persecution he had suffered for his religious beliefs, Richard Jowitt was too poor to make any such provision when he remarried.

When Richard died in 1696, his property, including grain, furniture, utensils, a heifer, two steers and '3 Fatt piggs,' amounted to a value of £82 5s, a not inconsiderable sum for the time. Unfortunately, his liabilities amounted to some £310. The main creditor, who had himself appointed administrator of Richard's estate, was Sir Gilbert Metcalfe, who had been Lord Mayor of York in 1695.[44] This may indicate that Richard had some connection to the wool business as Sir Gilbert was, until his death in 1698, Governor of the Merchant Venturers' Company, a York livery company involved in this trade. Ironically, Metcalfe's name is to be found on the Company's unsuccessful petition against the bill to make the rivers Aire and Calder navigable, a bill which was to have a profound effect upon the West Riding wool trade and thus the future prosperity of the Jowitt family.

There was nothing left for Richard's widow, the daughters of the first marriage, or the two surviving sons by his second marriage, Richard II (1694–1741) and Samuel (b. 1696), a third, Nathaniel (b. 1693) having died in infancy. Advised by the Friends' March 1697 Monthly Meeting not to contend the primacy of Sir Gilbert's claim on the estate, Mrs Jowitt did receive practical help from the Meeting with money to pay her rent. While life must have been hard for her, she found time to contribute to the Society, acting as the Gildersome representative to the Monthly Meeting between 1725 and 1727. She also received money from Robert Peart's legacy between 1727 and 1739 and an allowance from the Poor Fund between 1731 and 1739.[45]

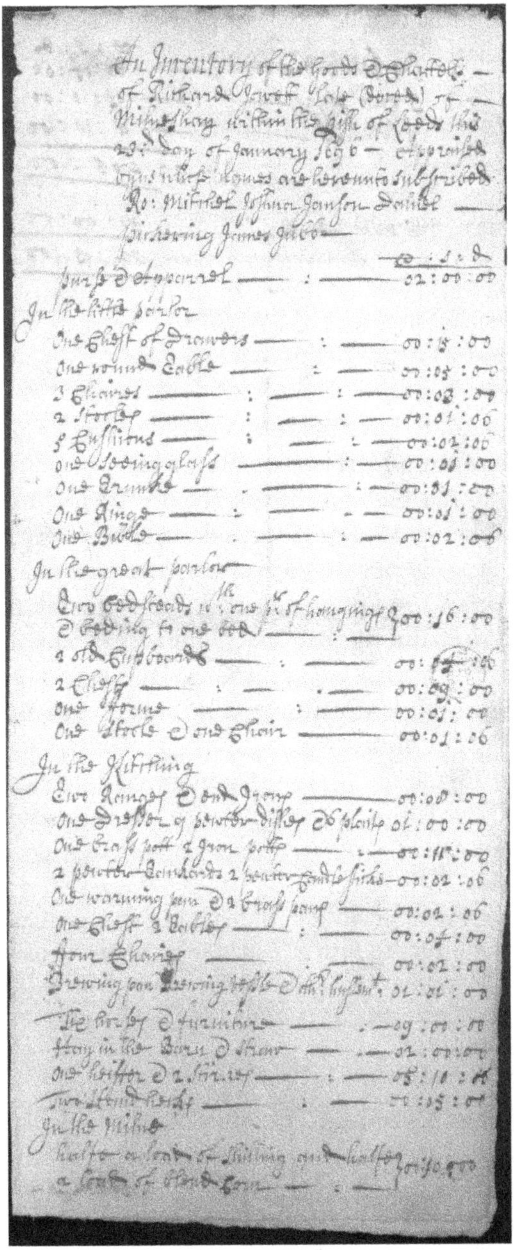

Fig. 2 Part of the inventory of Richard Jowitt I from the 1697 probate documents.
Reproduced by kind permission of the Borthwick Institute, University of York.

Richard's daughter Hannah also came under the care of the Friends. At her request, they placed her with Jonathan Merry and his wife Elizabeth (née Benson). The Friends also appointed Elizabeth Wilson and Mary Dowell to see what clothes she needed. The young Hannah soon incurred the displeasure of the Friends when it was reported that

> after a short time of her settlement [at the Merrys] she took her Cloth[e]s, & without acquainting her said Master or Dame or any other that took care of her, went privately away, & since hath shifted from place to place as her own inclination leads her; which Carriage is contrary to Truths' Orders & testyfyed against by Friends of this Meeting.[46]

Hannah also seems to have made an enemy of Mary Harryson, a fellow Quaker and wife of John Harryson of Beeston, who 'acquainted some of the Towns people that they ought not (or neede not) maintain Hannah Jawett by reason that she was not sprinkled at their Fount.'[47]

At the same time, it became known to the Leeds Meeting that Mr Merry's name was all too apt as he had a serious drink problem, not an uncommon affliction among the Leeds Quakers of the time.[48] He admitted 'keeping company with wicked men of the world' but promised to try and reform himself. Unfortunately, he was unequal to the task, and the Leeds Meeting cut him adrift.

The wretched Elizabeth Merry degenerated into 'a weake shattred wooman, & her words not consistant with Truth, or with themselves'. She attended the Leeds Preparative Meeting on 26 August 1702 and was

> discorced aboute her appearing in publick in our Meetings of Worshipp, & was desired for future to forbeare (as shee had beene privately admonished time after time), but instead of taking the advice of friends, she fell into extravigent reflections & uncomly speeches, wholy refusing the advice of this Meeting.[49]

The Meeting of 22 December 1703 unanimously declared its disunity with her, but the Leeds Friends relented and sent Abraham Jowitt (not, apparently, a relation) and Robert Walker to speak to her and she remained silent at the next monthly Meeting.[50]

Hannah Jowitt's sister Tabitha married one John Gott in 1704.[51] He seems to have been reasonably wealthy as he gave £2 towards the building of the new Leeds Meeting House.[52] Nonetheless, there is ample evidence that both Tabitha Gott (*née* Jowitt), like Tabitha Jowitt (*née* Hopwood), lived in straitened circumstances for many years, both receiving numerous subsistence payments from the Friends. It is extremely unlikely that the Friends would have made these payments if other members of the Jowitt family could have afforded to support them, so it would seem that the whole family remained in the clutches of extreme poverty in the late 1730s.

Richard II married Elizabeth Pearson of Holbeck in late 1720 or early 1721, but not until he had been forced to apologise to the Meeting for speaking to Elizabeth about the possibility of marriage before asking her father's permission: 'Friends showed their great dislike of such practices.'[53] Her father, James (b. 1673) had been one of the Quakers tried with Richard's father at the Leeds Quarter Sessions. The couple settled in Holbeck, then a separate village within the parish of Leeds. They had five sons and five daughters: John I, Elizabeth I (died aged 1), Richard III, Tabitha (died aged 3), Elizabeth II, Esther (died aged 8), Anna (died aged 2), Joseph I (died in the first year of life), Joseph II and Benjamin.

Richard II's son John (1721–1783) set up as a clothworker in Hunslet and married Ann Benson (1722–1802) at Gildersome on 18 June 1747. They settled in Pudsey but moved to Churwell in the 1760s. They had four sons, Richard III, John II, Joseph senior and Thomas, and three daughters, Mary, Elizabeth and Anna. On 6 January 1780, Elizabeth married Pim Nevins, a Leeds clothier whom many regarded as the finest woollen manufacturer in the country.[54] Nevins, who claimed descent from the ancient kings of Ireland, was a friend of Samuel Taylor Coleridge, who stayed with him while travelling to the Lake District and described Nevins as having 'woven into one web a Gentleman's delicacy and a Quaker's Honesty'.[55] Nevins also played host to a number of Seneca Indians from Buffalo, New York State, on a visit to England in 1818.[56] In 1785, Anna Jowitt married Samuel Birchall on 6 June 1785. He was a Leeds clothier and noted early collector of industrial tokens, the private coinage minted by industrialists to pay their workers. He wrote one of the first books on the subject.[57]

During the eighteenth century, Britain's growing empire created an increasing market for the products of her looms, although with one major setback, the American Revolutionary War. A statement by the manufacturers of Leeds, Woodhouse, Armley, Hunslet, Holbeck, Hightown,

Heckmondwike, Dewsbury and Batley expressed their growing concern about the effects of the war:

> These are to certify all whom it may concern, that from the total stagnation of the trade to North-America, great numbers of the labouring poor of this place are out of employ, and a great number that are but part employed; by which the distresses of the labouring poor are very much increased amongst us. And we the underwritten master manufacturers of woollen cloths already feel great inconveniences for want of that branch of trade as usual.[58]

One of the signatories was Joseph Jowitt senior of Hunslet, John II's younger brother.

Although two letters survive from J. Livelong of Potton, Bedfordshire, addressed to 'John Jowitt Clothier at Churwell' in 1775 and 1776,[59] it is clear that the business was even then more involved in selling raw wool than in manufacturing cloth. It was John II who finally ceased trade as a clothier and became purely a wool stapler. He moved the business from Churwell to Holbeck in the parish of Leeds. Although family legend in the nineteenth century credited him with being the first to move from Churwell (which was in the neighbouring parish of Batley) to Leeds,[60] his grandfather had himself been based in Holbeck.

The earliest business records we have for the family are contained in John II's ledger for 1775–1815.[61] The accounts show that wool was being purchased from East Anglia, the West Country, Hampshire and the London region and their customers (of whom there were seventy-five) were typically small clothiers in the Leeds area, such as William Brook and William Page of Morley, buying twelve or thirteen stones of wool per month. In 1775, credit terms varied from four weeks to six months and by 1776–7, they had extended further, reflecting the American situation. Joshua Wiliams of Ossett was given more than twelve months credit.[62]

John Jowitt II married Susanna Dickinson, a member of an old Gildersome Quaker family, on 31 August 1775. Her father Joseph, a white cloth manufacturer, had been delegated in 1756, with a Robert Walker, to buy the land for the first Gildersome Meeting House from one John Rayner. Like the Jowitts, the Dickinsons of Gildersome had been imprisoned in the early years of the Restoration for refusing to take the Oath of Allegiance and non-payment of tithes. One William Dickinson of Gilder-

Fig. 3 Joseph Firth of Toothill, by Maria Arthington.
Reproduced by kind permission of the
University of Leeds Library.

some was implicated in the Farnley Wood Plot of 1663, one of a number of incompetent northern insurrections against Charles II infiltrated by informers and *agents provocateurs*. Twenty-one of the twenty-six rebels condemned to death in the aftermath of the northern plots came from the clothing districts stretching from Leeds into the Pennines.[63]

In 1772, the Friends in the West Riding raised funds for a school and John Jowitt and John Elam, acting as cashiers for the Society, purchased a farm near Gildersome for that use.[64] The Jowitt family were to be much involved in educational provision over the next two hundred years.

John II's brother Joseph senior married Grace Firth at Paddock, near Huddersfield, on 30 May 1781. Perhaps in celebration of the happy union, Grace's brother Joseph Firth of Toothill, a wealthy Quaker, named one of his cows Jowitt. Other cattle were called Fruman, Haughs, Young Haughs, Belisha, Huddersfield and Strawberry.[65] The third Jowitt brother, Thomas, married Sarah Storer and moved to Nottingham.

The draft of a newspaper advertisement from this period (probably written in 1777, after the passing of a new act aimed at preventing fraud in the wool trade) shows that the firm was experiencing difficulties with sharp practices by wool collectors, in particular, 'winding'. Farmers, then as now, gave their flocks identifying marks, but in those days they were usually marked heavily with tar. Winding was the process of inflating the weight of wool by including this heavily marked wool,[66] or as mentioned in the advertisement, wool contaminated with dung or dirt:

23

— Wool —

To the Wool growers Wool Winders & Collectors of Wool in the Counties of Norfolk Suffolk and Glousc —

Whereas the Buyers of Wool in said Counties and Sorters thereof have been much Injured by the Unlawfull & Scandalous practice of winding within the fleece various kinds of Dirt frequently Sheeps Dung also fallen or Dead Sheeps Wool Lambs Wool unwashed Wool Clegg Locks &c to their great Deception & Loss —

This is to Certify All Wool Growers Wool Winders & Collectors of Wool in said Counties — that the Staplers and Dealers are Determined to open & examine the suspiciouse Fleeces in each Parcell on the Spot & to Prosecute According as the Law directs such Persons as may be detected in so fraudulent a practice.

By This timely Notice is given that none may be Ignorant of the late Act of Parliament which Inflicts four fold greater Penalties on false Winding than the former Law & a Short easy method of recovery also that Collectors of Wool are Responsible for Deceitfull Winding unless they do Prove & deliver up the real offender —[67]

The legal system, which had persecuted them less than a century earlier, had become the defender of their rights.

John I's eldest son Richard having died at the age of 16 in 1765, his other two sons, John II and Joseph, continued the family wool business after their father's death in 1783, forming a partnership with Samuel Birchall of Hunslet, who married their sister Anna in 1785. It was under the leadership of John II (Jno. Jr. as he usually signed himself, just as his grandson was to many years later) that the company started to grow from a small local concern into a substantial business. It is clear that John II, as well as trading wool, provided basic banking services (particularly loans) to a number of personal and business acquaintances, including Nevins & Gatliff, the Hunslet woollen mill co-owned by his brother-in-law Pim Nevins, which owed £2000 in 1806.[68] Such lending by merchants was not uncommon and was a service undertaken by other businessmen in that period, including the Jowitts' relations, the Wilsons of Stang End and of High Wray.[69] Moreover, this was clearly a service to the community rather than a commercial operation, sometimes actually run at a loss.

John II charged interest of 5% per annum over many years, including between 1794 and 1801, during which period inflation rose by about 55%,[70] while 5% compound interest for the same period would have amounted to 48%.

As with many areas of manufacturing, the wool trade was transformed during the course of the eighteenth century. This was not just as the result of new carding, spinning and weaving machines but also wide-ranging organisational changes. Larger scale production encouraged specialisation. Whereas earlier generations of clothiers might have sourced their wool locally, by the second half of the eighteenth century they were reliant on increasingly large wool staplers with nationwide, and indeed international, connections. This also allowed clothiers to produce a wider range of cloths, especially the more expensive worsted and 'mixed' cloths which required finer, long-staple wool than was widely available domestically.

By the end of the eighteenth century, the wool trade had shifted significantly from the West of England to the West Riding. The impression is sometimes given that this shift occurred because the Yorkshire workers were less resistant to the introduction of new machinery than those of the West Country. Quite the opposite was the case at first. While it is true that there were riots against mechanisation in the West Country, there were also riots in Leeds and Bradford. The introduction of power-spinning in Leeds provoked a riot in 1794 and workers destroyed a Leeds mill in 1799.[71] William Hirst, a successful Leeds woollen manufacturer and cloth-finisher, remarked in 1844:

> About sixty years ago an attempt was made to introduce machinery for finishing cloth, both in the West of England and in Yorkshire. The workmen raised the most violent opposition to it, and after a severe struggle, the masters in Yorkshire were obliged to abandon the attempt; while in the West of England, they succeeded. Thus they had a double advantage, for all their goods were manufactured and finished under their own care, while those in Yorkshire were manufactured in various parts, and to sell in the balk [bulk] state. They were then sent out to be finished... [72]

Hirst, who says that innovation has been 'the salvation of the woollen trade in Yorkshire,' tells us that he felt his life was under threat from his workmen every time he introduced a new piece of machinery.[73]

John Jowitt & Son

... we have concluded by mutual consent to dissolve our present Partnership which we think will be effected within about three months of this time. I hope our doing so will be advantageous to all parties, as we are now too many for one house, our Jno. and Josh. having each a son who are training and likely in Business; each the managements of our connections in Norfk. Sufk. etc. we have agreed amongst ourselves that you be continued to do Business as usual, for and to Accot. of our Jno. Jowitt /say Jno. Jowitt & son/ if agreeable to you...[74]

On 10 June 1802, the partnership between the brothers was dissolved, with creditors advised to seek payment from any of the previous partners. John II formed John Jowitt & Son while Joseph remained in partnership with Samuel, trading as Jowitt & Birchall. There is no mention at this stage of Samuel's sons going into the business, although his eldest, Samuel Jowitt Birchall, was fourteen by this time. The split seems to have been entirely amicable, though the difficulties of dividing up the assets and re-establishing links with customers and suppliers must have been daunting, and point to a deep-felt need to reorganise. For many years the two firms had offices almost next door to each other in Albion Street, Leeds, and both also had warehouses in Hunslet Lane. When it was placed on the market many years later, John Jowitt & Son's property in Hunslet Lane, then a very desirable part of town, was described as

Two good Dwelling-Houses... a large Warehouse about Thirty five Yards Long, Eleven Yards wide, and Three Stories high, another Warehouse about Fourteen Yards long, Seven Yards wide, and Three Stories high, also a Counting-House and Packing-Shop with Chambers over them, Stabling for Seven Horses, &c. &c. and about Two Acres of Land adjoining, in Garden, Plantation and Field, with an open Front to Hunslet-Lane of One Hundred and Fifty yards.

It was said to be particularly suitable for a merchant 'wishing to set up Machinery for manufacturing or dressing Cloth'.[75]

At the time of the dissolution of the partnership between John Jowitt II, Joseph Jowitt and Samuel Birchall, John II had £6,469 8s 9d capital plus £1,544 5s 6d interest invested in the partnership, Joseph had £6,744

Fig. 4 John Jowitt II (1750–1814).

8s 4d plus £2,344 15s 8d and Samuel had £735 9s 9d plus £754 19s 6d (having recently withdrawn most of his capital).[76]

JOWITT & BIRCHALL AND JOHN JOWITT & CO.

From this point on, Jowitt & Birchall and what was later to become Robt. Jowitt & Sons, although both dealing in wool in Leeds, seem to have had no commercial contact, other than a loan of £500 made by John II to Samuel Birchall on 21 October 1802, which was repaid with interest over the next few years.[77]

Joseph Jowitt died in 1803 and was succeeded by his sons, John (known as John senior – not to be confused with Johns I and II), Thomas and Joseph (for a fuller discussion of Joseph Jowitt's descendants, including a Lord Chancellor and an eccentric millionaire, see Appendix XIII). Few records survive of the Jowitt & Burchall business, but Samuel's son clearly prospered because in 1813, at the age of twenty-five, he was able to buy

27

BEWARE OF THE FEAST

♂ **John Jowitt I**
b. 10 May 1721 at Holbeck
m. 18 Jun 1747 Ann Benson (1722 - 1802)
d. 1783 at Churwell, aged 62

♂ **Richard Jowitt III**
b. 14 Jul 1748 at Pudsey
d. 18 Mar 1765, aged 16

♂ **John Jowitt II**
b. 04 Aug 1750 at Churwell
m. 31 Aug 1775 Susanna Dickinson (1752 - 1819)
d. 15 Dec 1814 at Leeds, aged 64

♀ **Mary Jowitt**
b. 09 Nov 1752 at Pudsey
d. 03 Apr 1754, aged 1

♀ **Elizabeth Jowitt**
b. 07 Jun 1755 at Pudsey
m. 06 Jan 1780 Pim Nevins (1756 - 1834)
d. 01 Mar 1802, aged 46

♂ **John Jowitt**
b. 08 Aug 1776 at Churwell

♀ **Ann Jowitt**
b. 1777
d. 01 Feb 1837
diary 1847, aged 60

♀ **Elizabeth Jowitt** – – – – – – – – – –
b. 1779
m. 27 Jul 1803 Isaac Crewdson (1780 - 1844)
d. 1855, aged 76

♂ **John Jowitt**
b. 22 Jul 1782 at Leeds

♂ **Richard Jowitt**
b. 30 Jul 1783 at Leeds

♂ **Robert Jowitt** – – – – – – – – – –
b. 1784 at Churwell
m. 08 Feb 1810 Rachel Crewdson (1782 - 1856)
d. 19 Dec 1862 at Leeds, aged 78

♀ **Mary Jowitt** – – – – – – – – – –
b. 1786 at Leeds, England
m. 1808 Isaac Wilson (1784 - 1844)
d. 02 May 1846 at Kendal, England, aged 60

♀ **Susannah Jowitt**
b. 26 May 1787 at Leeds
d. 1813 at Leeds, aged 26

♀ **Rachael Jowitt** – – – – – – – – – –
b. 05 May 1791 at Leeds
m. 12 Oct 1815 Joseph Crewdson (1784 - 1844)
d. 03 Jul 1826, aged 35

♀ **Hannah Jowitt**
b. 28 Mar 1793 at Millshaw, Leeds
m. 12 Oct 1815 William Wilson
d. 15 Aug 1875 at Kendal, aged 82

♂ **Joseph Jowitt sr** – – – – – – – – – –
b. 01 Oct 1757 at Pudsey
m. 30 May 1781 Grace Firth (1758 - 1846)
d. 17 Mar 1803 at Leeds, aged 45

♂ **Thomas Jowitt** – – – – – – – – – –
b. 29 Jan 1760 at Pudsey
m. 13 May 1784 Sarah Storer (- 1779)
d. 22 Sep 1789, aged 29

♀ **Anna Jowitt** – – – – – – – – – –
b. 14 Mar 1765 at Churwell
m. 06 Jun 1785 Samuel Birchall (1761 - 1814)
d. 11 Feb 1793 at Leeds, aged 27

Fig. 5 Carlton House, John Jowitt II's house on Woodhouse Lane.

the magnificent Springfield House following the bankruptcy of Thomas Livesey, clothdresser. The house is now part of the University of Leeds. The partnership was then renamed Thomas and John Jowitt & Co. Samuel Jowitt Birchall (who became a director of the Leeds and Yorkshire Assurance Company)[78] left the partnership in 1821[79] and Thomas and Joseph retired on 30 December 1837,[80] leaving John Jowitt senior to keep trading as John Jowitt & Co.

JOHN JOWITT II AND HIS FAMILY

John Jowitt II's business continued to prosper through the turbulent years of the early nineteenth century. When, after a short peace, Britain was plunged into another war with France in 1806, it destroyed the fortunes of many wool merchants, including Richard Bramley who had twice served as Mayor of Leeds. Bramley had been one of the richest Leeds wool merchants and had lived in great style, building an elegant villa on Woodhouse Lane on high ground to the north of the town. He named it Carlton House after the Prince Regent's opulent palace on the Mall in London. In the wake of Bramley's business failure, the house was bought by John Jowitt II whose lack of extravagance had no doubt saved his own business from a similar fate. It is hard to imagine anyone who had less in common with the Prince Regent than John II, but he retained the name Carlton House and there lived for the rest of his life in some comfort, surrounded by about 2.6 acres of paddock and pleasure grounds[81] in a prime location, removed a little from the increasingly polluted air of industrial Leeds.

Robert Jowitt

Although John II had four sons (two Johns, Richard and Robert), only Robert survived into maturity and entered the family firm. Robert married Rachel Crewdson, daughter of Thomas and Cicely Crewdson (*née* Dillworth) of Kendal, on 8 February 1810. Rachel's marriage settlement (an arrangement whereby some or all of her assets were put into trust before her wedding – see Appendix II) shows her to have had a fortune of over £3,500 (about £210,000 at current values). There can be no doubt that it was a very happy marriage. Thirty-seven years later, Robert was to write,

> ... how greatly blessed we have been in our dear children, & in being permitted to believe they are the Lord's children, oh what a favour to one so unworthy as the writer feels & to have yet spared to me, the partner of my joys and sorrows, my counsellor & comforter, how can I be sufficiently grateful for such a blessing?[82]

John Jowitt II's Six Daughters

John II also had six daughters, Ann, Elizabeth, Mary, Susanna, Rachael[83] and Hannah. Elizabeth Jowitt married Isaac Crewdson, Rachel's brother, in 1803, and Rachael Jowitt married Isaac's brother Joseph in 1815, at which point there was a Rachael Crewdson *née* Jowitt and a Rachel Jowitt *née* Crewdson. In 1808 Mary Jowitt married Isaac Wilson, son of John and Sarah Wilson (*née* Dillworth) of Kendal in 1808 and her sister Hannah married Isaac Wilson's brother William in 1815. Isaac and William Wilson's grandmother was the indomitable Rachel Wilson, a Quaker minster from the age of eighteen, who travelled to visit members of the Society of Friends in Ireland in 1754. She was at that time described as 'an able skilful Minister, and deeply read in the Mysteries of the Kingdom'.[84] In July 1768 she set off from Kendal to America on a similar mission, arriving within the Capes of Delaware on 14 October, travelling to New Castle (DE), Wilmington (DE) and Philadelphia, where she preached against the slave trade, 'believing it never was intended for us to traffick

Fig. 6 'Modesty is not only an ornament, but also a guard to virtue...' Robert Jowitt's writing practice, dated 19 November 1790. The copperplate handwriting is remarkably fine for a six-year-old. Like most Quakers, the Jowitts placed great importance upon education.

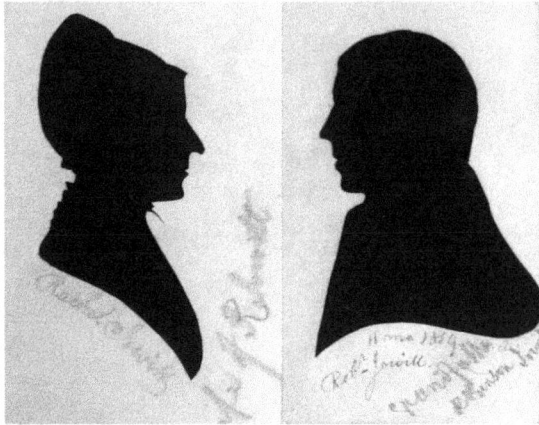

Fig. 7 Rachel and Robert Jowitt, 1819, by Maria Arthington *née* Jowitt.
Reproduced by permission of the University of Leeds Library.

with any part of the human species.' In all, she travelled several thousand miles in America, even accepting an invitation to preach to Princeton students.[85]

Five Jowitts in a single generation had married people with close Kendal connections and very close bloodlines. The two remaining sisters, Ann and Susanna, died spinsters.

Isaac and Joseph Crewdson were wealthy wool and silk merchants in Manchester, and in 1788 their father had co-founded, with Joseph Maude and Christopher Wilson, what was variously known as Maude, Wilson & Crewdson's Bank, Wakefield Crewdson and the Kendal Bank. Like many other banks, it issued its own bank notes. Unlike most of them, it survived the many banking crises, merging with the Bank of Liverpool (later Martins Bank, now Barclays) in 1893. The Crewdsons, like the Jowitts were married into a number of important families from Kendal and thereabouts, including the Wilsons, Dillworths, Braithwaites, Birkbecks, and Bensons. This extended Quaker connection was to have a profound impact on the Jowitt family, both socially and financially. It was also almost inevitable since the Quakers strongly disapproved of mixed marriage and, indeed, regularly expelled members for marrying outside the Society.[86] There is no doubt that the Jowitts took these family ties very seriously. A brother-in-law is always referred to as simply 'brother', cousins

32

Fig 8 Kendal Bank, 9 Highgate, Kendal, the successor to the bank formed by Joseph Maude, Christopher Wilson and Thomas Crewdson on 1 January 1788. The Jowitts were related to both the Wilsons and the Crewdsons.

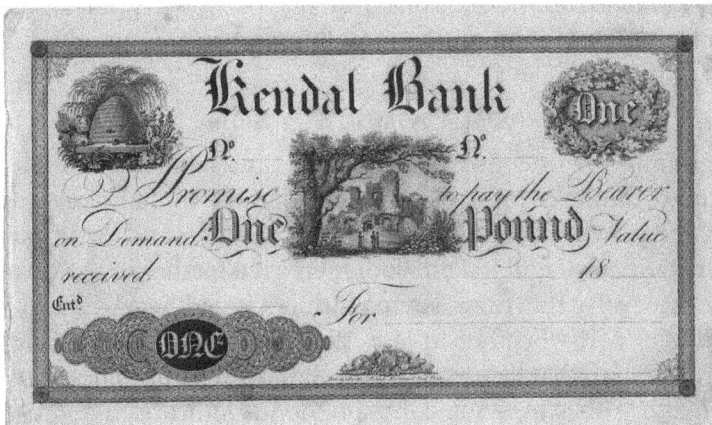

Fig. 9 A Kendal banknote. Photograph reproduced by kind permission of Dix Noonan Webb.

33

Fig. 10 Pim Nevins, 1812, by Maria Arthington *née* Jowitt.
Reproduced with the permission of the University of Leeds Library.

are always addressed as 'cousin', aunts as 'aunt' and so forth. The cousins were safe partners in both marriage and business, and constant guests and hosts at tea or supper. When in London, the Jowitts would stay with the Forsters or the Howards, when in Manchester with the Crewdsons, when in Kendal with the Bensons. Equally, the Jowitts' houses were always full of visiting relations, there on business or pleasure.

JOHN JOWITT & SON

Although John Jowitt & Son's activity still consisted largely of trading in domestic wool, the company had a considerable trade in foreign, especially Spanish, wool in the late eighteenth century,[87] and the importation of German, Spanish and Italian wool increased further in the early nineteenth century. By this time, some clothiers relied heavily on imported wool and John II's brother-in-law, Pim Nevins, told a parliamentary committee in 1816 that he used only foreign wool – Spanish, German and some French.[88] The surviving correspondence indicates that most of the Jowitts' suppliers in the early nineteenth century were based in East Anglia, Essex, Kent, Hampshire and Hertfordshire. As might be expected, the firm tried hard to squeeze the best prices out of suppliers: 'If Business can be done of reasonable terms...,'[89] 'sorry we cannot comply with the Prices...,'[90] etc. Given the remoteness of the suppliers in those pre-railway days, much of the purchasing was necessarily based upon sight of samples or previous consignments which might not be matched. One supplier was told bluntly, 'last Super was not equal to the former parcell I shall expect the prices to be lower...,'[91] another was requested to send someone 'who is a judge of wool so that he'll see how inferior the latest batch is...'[92]

Transport was one of the biggest problems facing the firm. Although the roads had been much improved during the eighteenth century, road transport was both slow and expensive. The canals could also be slow. John Jowitt & Son shipped much of their wool from the southeast of England by sea, from London or Dover, although sea transport could be hazardous. One consignment in 1803, from George Finch & Co. in Dover, aboard the vessel *Hope* bound for Selby, was driven off-course to South Shields, where it foundered. Fortunately, in this case, the firm's warehouseman, who had been sent to examine the cargo, found that it was little harmed, 'except for a few packs, a little wet, but not materially so'.[93] However, wool could suffer water damage just as easily on road journeys. Another shipment to Selby, from Elsted Warnham in Charing, Kent, arrived in a very wet and damaged state. Jowitts claimed that this must have happened on the journey from Charing to London and deducted an appropriate amount from the bill, advising the supplier to seek restitution from his carrier.[94]

On 15 December 1814, John Jowitt II was found dead in his garden at Carlton House. He was over sixty-four, by no means young by the standards of the time, but his sudden death shook his son Robert so much that he was still lamenting his loss seventeen years later. On 15 December 1831, John junior recorded that his father Robert had 'just been giving us an interesting account of the event, & his feelings on it – as one would expect, they were awful, & the mention of them, seems even now almost to make him shudder.'[95]

At the time of his death, John II was worth £30,000.[96] This is the equivalent of about £1.8 million at current values.[97]

From John Jowitt & Son to
Robt. Jowitt & Son

> Mr Jowitt, a wool-stapler, states that with respect to English [short staple] wool, he has found a very great difference in the quality; that his customers are now obliged to get foreign wool for what they used to make of English wool, partly because the latter is deteriorated, and partly because having been accustomed to use finer, they found it suits their purpose better.[98]

With growing customer demand for foreign wool, it was inevitable that John Jowitt & Son would become increasingly involved in the international trade. There were, however, conflicting interests in the British economy, with native producers of raw wool wishing to protect their trade from foreign competition and British producers of cloth arguing that access to foreign wool was essential if they were to compete with foreign manufacturers. This led to considerable uncertainty in the British wool trade, as different vested interests jostled for position, sending levies on wool imports up and down, from 5d per lb in 1819 to 3d in 1825.[99] Foreign wool imports for home consumption fluctuated greatly between 1820 and 1831 (see Table 1) due to changes in both demand and levies.

Year	Tonnes	Year	Tonnes
1820	3489	1826	8105
1821	7211	1827	12675
1822	7374	1828	14076
1823	8522	1829	10258
1824	10884	1830	14299
1825	18643	1831	13472

Table 1: Wool imported into Britain for domestic consumption.[100] Original figures in pounds have been converted to metric tonnes.

At the same time, foreign tariffs on imported British cloth could have a significant effect on exports. The United States increased tariffs in 1814, in the wake of the War of 1812, and protectionist pressures led to further increases in 1824 and 1828.

Profits in the wool trade were variable and losses were not uncommon. Robt. Jowitt & Son made a profit of £5,085 19s 3d in 1833 (the year in which it took that name), falling to £1,164 8s in 1834 and £761 4s 9d in 1835. After a partial recovery to £2,620 2s 9d in 1836, the company made a loss of £6,242 10s 9d in 1837.[101] Moreover, the long credit terms which the Jowitts were obliged to give their customers (often as much as ten months, though they were able to charge interest on this)[102] meant that the partners had to keep most of their capital in the firm. For instance, of Robert's total wealth of £21, 841 4s 10d in 1835, £20, 547 11s 10d was invested in the company.[103] Nevertheless, Robert managed to diversify his investment portfolio, giving him some protection from fluctuations in the wool price. As well as a few private properties in Leeds, he had shares in Leeds baths, the market, and the local gas and water companies.[104] He also invested in transport projects such as the Stockton & Darlington and Liverpool & Manchester railways and the Humber Union Steam Company. Given his many charitable donations (of which more later), many of Robert's investments were probably made with the intention of improving the local community rather than making a profit. At the peak of the 1830s railway mania, 8% of GDP was invested in infrastructure projects. One modern writer put that into perspective by calculating that the same percentage of the US economy today would be worth about $2 trillion.[105] Unlike other bubbles, including the revived enthusiasm for the railways of the 1840s, this bubble did not burst and returns to investors were good. Robert also invested £300 in the Victoria Bridge linking Leeds and Holbeck, built 1837–8, at a total cost of about £8,000, and lost every penny.[106]

Although in most respects benevolent and liberal-minded, Robert retained what might be thought to be a puritanical streak from his Quaker background. In 1834, he addressed a notice to the inhabitants of Woodhouse, especially those who 'have found it impossible, *without the help of others,* to obtain the necessaries of life,' warning them to 'BEWARE OF THE FEAST.'[107] While not wishing 'to encroach upon the rational enjoyments of the labouring classes, or to lessen their comforts,' he argued that the Feast commenced in Sabbath-breaking, that it wasted time and money and led to drunkenness and undesirable behaviour. Many middle-class temperance supporters felt that alcohol in moderation was acceptable but that the lower orders drank immoderately and should be dissuaded from doing so for their own sakes. Robert, a fervent supporter of the British and Foreign Temperance Society, went further:

> If, amongst those of a higher class, we could see how often the habit of in-temperance mars the comfort of many otherwise happy families, robbing them of that peace and enjoyment in each other's society, which ought to be their mutual solace through the journey of life,—how the faculties are benumbed, the finer sensibilities of the mind destroyed, and in how many various forms of disease the effects of intemperance are visible, and often, when not visible for a time, are powerfully at work, undermining the constitution, and eventually bringing on premature old age and even death itself: could we ascertain the number of deaths of which intemper-ance is, not the *remote,* but the *immediate* cause, in cases of drunken cart drivers, waggoners, and coachmen; could we number the lives lost at sea, and upon our rivers and canals, and in a variety of of other ways from the same cause: if to these results of the *intemperate use* of liquor, we could add the baneful effects of what is called "drinking in moderation," or the "prudent use" of spiritous liquors, and observe how the pernicious habit is *slowly and unsuspectedly, but gradually and certainly* bringing on a num-berless train of evils, and inflicting serious injury upon the constitutions of many who maintain the character of sober men, and who rarely if ever drink to intoxication:—was it possible to see at a glance the various evils thus briefly hinted at, and which form but a part of those actually flowing from the same prolific source,—I think many would unite with me in the conclusion, that of all the evils which bring upon themselves, INTEMPER-ANCE is the source of most misery and crime.[108]

Despite his strictures on the dangers of drinking, even in moderation, we find that Robert had a liquor account in his private ledger.[109] It is ad-mittedly a fairly modest account which, at its height consists of port and sherry to the value of £8 9s (about £770 at modern values) for a whole year, but it is surprising nonetheless.

Robert was also an extremely active philanthropist, as is shown by the many pages of his personal ledgers devoted to charitable donations.[110] As well as regular gifts to the Leeds Anti-Slavery Society (of which he was Chairman), the Animal Humane Society, various Bible societies, schools and Leeds General Infirmary, many individual donations are recorded to 'A man who had lost a horse,' 'An American,' 'Widow Stacey,' 'Widow Nettleton,' 'Destitute female,' 'J. J. Thompson of Georgia,' 'Blind man,' 'French man,' etc. His spending on books generally reflected similar con-

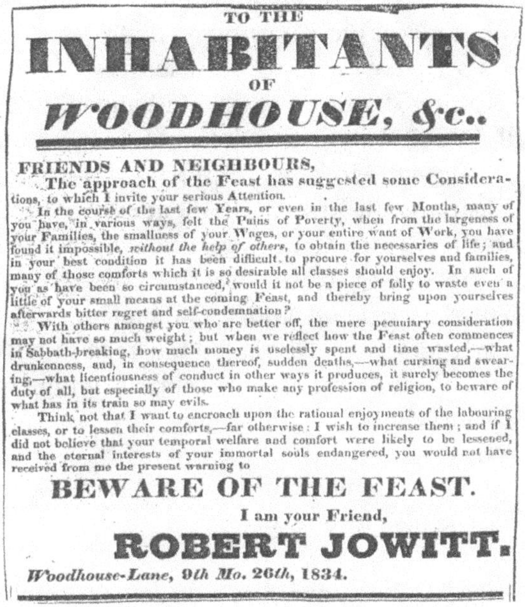

TO THE

INHABITANTS

OF

WOODHOUSE, &c..

FRIENDS AND NEIGHBOURS,

The approach of the Feast has suggested some Considerations, to which I invite your serious Attention.

In the course of the last few Years, or even in the last few Months, many of you have, in various ways, felt the Pains of Poverty, when from the largeness of your Families, the smallness of your Wages, or your entire want of Work, you have found it impossible, *without the help of others*, to obtain the necessaries of life; and in your best condition it has been difficult to procure for yourselves and families, many of those comforts which it is so desirable all classes should enjoy. In such of you as have been so circumstanced, would it not be a piece of folly to waste even a little of your small means at the coming Feast, and thereby bring upon yourselves afterwards bitter regret and self-condemnation?

With others amongst you who are better off, the mere pecuniary consideration may not have so much weight; but when we reflect how the Feast often commences in Sabbath-breaking, how much money is uselessly spent and time wasted,—what drunkenness, and, in consequence thereof, sudden deaths,—what cursing and swearing,—what licentiousness of conduct in other ways it produces, it surely becomes the duty of all, but especially of those who make any profession of religion, to beware of what has in its train so may evils.

Think not that I want to encroach upon the rational enjoyments of the labouring classes, or to lessen their comforts,—far otherwise: I wish to increase them; and if I did not believe that your temporal welfare and comfort were likely to be lessened, and the eternal interests of your immortal souls endangered, you would not have received from me the present warning to

BEWARE OF THE FEAST.

I am your Friend,

ROBERT JOWITT.

Woodhouse-Lane, 9th Mo. 26th, 1834.

Fig. 11 Robert Jowitt's warning to the inhabitants of Woodhouse in 1834.

cerns: the works of George Pilkington, a campaigner against slavery in Brazil, temperance and anti-slavery tracts, a subscription to the *British Emancipator* and three copies of a publication on Lutheran persecutions.

He was particularly active in the educational field, teaching at the Friends' Adult Day School in Leeds and being a member of the Country Committee (as distinct from the London Committee) of Ackworth School, near Pontefract.

In addition, Robert was a minister at the Brighouse Monthly Meeting from 1821.[111] Despite being in constant pain from rheumatism, he went on a mission to the Quakers in Ireland in 1832, rather as his wife's grandmother, Rachel Wilson, had done in 1754, arriving in Dublin in the middle of a cholera epidemic (which appeared in other parts of the country too, including Belfast). On landing in Dublin, he was sorry to observe

> so much ragged poverty amongst the poor; many of the women & children without shoes or stockings.[112]

From 18 February to 30 April, he followed a gruelling schedule, travelling (amongst other places) to Kingstown, Wicklow, Wexford, Ballitore (where there was a Quaker school), Enniscorthy, Cooladerry, Forrest, Ross, Kilconna, Carlow, Ballitore again, Rathangan, Edenderry, Mountrath, Ballinakill, Knock Ballymaken, Roscrea, Beir, Moate, Ballymoney, Cootehill, Lower Grange, Moyallon, Rathfriland, Lurgan, Richhill, back to Moyallon, Hillsborough, Lisburn, Ballenderry, Antrim (including a brief visit to see the Giant's Causeway), Ballynacree, Belfast, Lisburn again, Waterford (for the Quarterly Meeting), Drogheda, back to Dublin, Mountmellick, Roscrea and Knock Ballymaken, Carlow, Dublin yet again (for the Yearly Meeting). He visited fourteen ministers (more than half of them women) and a number of schools, and came away with a very favourable impression:

> The character of the Irish appears to me, warm hearted, generous & kind, very frank & open in their manners...

From 1845 to 1860 Robert served as a director of the Leeds and Yorkshire Assurance Company[113] and as one of the early directors of the Friends Provident Institution,[114] founded as a friendly society by his fellow Quakers, Joseph Rowntree and Samuel Tuke in 1832.

He was Treasurer of the Fund for the Relief of the Unemployed Poor, a member of the Committee of the Ilkley Bath Charitable Institution and of the British and Foreign School Society, Chairman of the Leeds Anti-Slavery Society, President of the Leeds Benevolent or Strangers' Friend Society, and a Trustee of the Leeds Guardian Asylum and Female Penitentiary (and, for his pains, had garments stolen by two escaping inmates of this female institution).[115] He was still serving on these last two institutions in the final year of his life.

Robert and Rachel lived for many years at what had formerly been his father's home, Carlton House. After Robert's death, the house and part of the land were sold to the Friends who built an attractive neoclassical Meeting House there, appropriately designed by Edward Birchall, brother-in-law of Robert's grand-daughter Emily. Later it was used as the local BBC offices, more recently becoming part of Leeds Metropolitan University. They had five daughters, Elizabeth, Susanna, Mary Ann, Rachel and Esther, and two sons, John junior and Robert Crewdson (RCJ).

Portraits of Robert show him looking rather dour, but whether this was due to a rather austere Quaker streak or from physical suffering it is hard to say. His diaries show that he had rheumatism and numerous other complaints – so numerous, indeed, that they might lead one to suspect severe hypochondria. When he sprained his foot in 1822, it resulted in an infected blister and pains in the head, throat, back and bowels which kept him off work for eleven weeks.[116] His rheumatism was undoubtedly real and led him to consult the eminent London physiologist, Sir Benjamin Brodie, author of *Diseases of Joints,* who referred him to the equally eminent surgeon, Sir Astley Cooper. John junior noted, 'Sir A. Cooper thinks they can cure him, but must make him worse, before they make him better...'[117] which, to the layman, sounds very much as though they had no idea what they were dealing with. Despite the rather pained appearance shown in his portrait, Robert was clearly a kind and doting father with a great sense of fun. For instance, on 4 September 1823 he took John junior to see a balloon ascent:

Mr W. Sadler ascended from the Colo[u]red Cloth Hall Yard with a balloon 42 feet by 33 ft. diameter, but it was only inflated about 2/3, Jno. & I saw him from the roof of our warehouse & watched him for 14 minutes when we lost sight of him...[118]

and on the following day:

Saw Green ascend from the yard of the White Cloth Hall, the day was very fine & the wind moderate – the ascend [sic] was a beautiful sight to behold, I saw the Balloon 25 minutes after it left the Yard.

Robert and Rachel's daughter Mary Ann married Edward Whitwell of Kendal, a carpet manufacturer. Their grand-daughter Louisa Crommelin Roberta Jowitt Whitwell became Duchess of Bedford on 27 August 1940 when her husband Hastings William Sackville Russell acceded to the title of twelfth duke.

Robert had an enquiring mind and, even when travelling on Quaker business, as he often did, he would make time to investigate natural and man-made objects of interest and note them in his diary. Perhaps because he travelled so much by road, he was particularly interested in bridges, commenting on the iron bridge at Wykeham erected by Rich-

Fig. 12 Drawing of the obelisk next to the Howth lighthouse, Dublin bay, from Robert Jowitt's diary, 1832.

ard Langley in 1802.[119] He was equally taken by the obelisk next to the Howth lighthouse in Dublin bay.[120] Having attended Society Meetings at Scarborough and Filey in 1842 and parted company with J. J. Gurney and Isaac Crewdson, Joshua Priestman and Robert made a detour to the Kirkdale Hyena Cave, where the fossilised remains of many animals including hyenas and a hippopotamus had been found three years before.

John Jowitt Junior
(1811–1888)

How comes it that with a few exceptions, our womenkind are so ordinary?
John Jowitt junior, 1831[121]

John junior (as he continued to call himself for some forty years after his grandfather's death, no doubt to avoid confusion with his great-uncle Joseph's son John)[122] was sent to a local day school run by a Mr Mercer[123] and received drawing lessons from Alphonse Dousseau, an artist now mostly remembered for his panorama of Leeds.[124] John's lively sketches and watercolours remain an interesting record of his travels.

At the age of twelve, John junior was sent to Josiah Forster's school in Tottenham near London. His parents did not despatch their son to boarding school without some pangs of regret. Robert recorded in his diary:

> This morning we have parted from our son John, who went this morning per coach to Manchester, in order to proceed tomorrow to Tottenham with his cousins J[ohn] J[owitt] Wilson and Geo. Benson. It has cost us some feeling, but tho the thought occurred that we might never see him again in this world, it did not distress me; for conscious that we were designing his good in this separation; I felt able with humble confidence to command him to that Protection which can alone be his safeguard, at home or absent from us, he is nearly 13 years old, & has never I think been separated from us a longer term than about 5 weeks...[125]

Robert was probably thinking of his own frailty, although he lived for another twenty-nine years, but the risks to a child attending boarding school in those days should not be under-estimated – at least eighty-eight children contracted typhus at Ackworth School in 1831 and at least one of them died from it.[126]

Like almost everyone the Jowitts knew, Josiah Forster was one of their extended Quaker family, being married to Rachel Wilson, sister of Isaac Wilson who had married Robert's sister Mary in 1808. Like Robert, Forster served on the Country Committee of Ackworth School. His Tottenham establishment was considered at the time to be the best Quaker school in the country, and it was said that John junior was 'much better

Fig. 13 John Jowitt junior (1811–1888).

furnished, mentally, than many a graduate of Oxford or Cambridge'.[127] Nonetheless, it seems clear that he had not really enjoyed his time at Tottenham because on 10 August 1835 he wrote,

Poor Bob [John's younger brother, Robert Crewdson Jowitt] went back to school today – my heart ached for him when I thought of Tottenham but I hope York is better & I *believe* that he is different to what I was – oh when I think of that, & God's amazing mercy, how can I be such an ingrate![128]

When Josiah Forster closed his school in 1826, John Junior's formal education finished, and he joined the family firm before his fifteenth birthday. For many years he felt the need to make up for the shortness of his education by rising early to study French, German or the classics. He would read the New Testament in Greek.[129] It is also clear that, although the family business was thriving, he found it lacking in dynamism or, as he put it, 'sleepy'.[130]

Descriptions of the young John make it clear that he himself was a bundle of energy with 'an abundance of thick black hair, very penetrating dark eyes, a springy elastic step, great rapidity in all his movements, and conveying a general sense of extreme brightness and vigour both of

♂ **Robert Jowitt**
b. 1784 at Churwell
m. 08 Feb 1810 Rachel Crewdson
 (1782 - 1856)
d. 19 Dec 1862 at Leeds, aged 78

♂ **John Jowitt jr**
b. 15 Sep 1811 at Kendal, England
m. 05 May 1836 Deborah Benson
 (1813 - 1893)
d. 30 Dec 1888 at Leeds, aged 77

♀ **Elizabeth Jowitt**
b. 1816

♀ **Susannah Jowitt**
b. 1818 at Leeds
d. 1859, aged 41

♀ **Mary Ann Jowitt**
b. 1819 at Leeds
m. 08 Jul 1841 Edward Whitwell (1817 -
 1893)
d. 1878 at Kendal, aged 59

♀ **Rachael Jowitt**
b. 1821 at Leeds
m. 20 Apr 1817 Andrew Reed

♂ **Robert Crewdson Jowitt**
b. 1821
d. 03 Oct 1847 at Leeds, aged 26

♀ **Esther Maria Jowitt**
b. 1826

♀ **Susan Maria Jowitt**
b. 04 Aug 1837 at Leeds
m. 26 Apr 1860 Theodore Howard

♂ **Robert Benson Jowitt** – – – – – – –
b. 1839 at Leeds
m. 06 Sep 1865 Caroline McCulloch
 (1844 - 1921)
d. 1914 at Tunbridge Wells, aged 75

♀ **Rachel Elizabeth Jowitt**
b. 19 Oct 1841 at Leeds
m. 14 Aug 1864 Theodore Crewdson
 (1835 -)
d. 28 Jan 1880 at Fernacre. Alderley
 Edge, aged 38

♀ **Anna Dora Jowitt** – – – – – – – – –
b. 1844 at Leeds
m. 1865 David Howard (1839 - 1916)
d. 1935, aged 91

♀ **Emily Jowitt**
b. 1853 at Leeds
m. 22 Jan 1873 John Dearman Birchall
 (1828 - 1897)
d. 1884, aged 31

♀ **Florence Jowitt**
b. 1855 at Leeds
m. 1889 Arthur Paine Baines
d. 19 Aug 1927 at Leeds, aged 72

♀ **Rachel Whitwell**
b. 1845

♀ **Mary Whitwell**
b. 1846

♀ **Frances Whitwell**
b. 1852

♂ **Edward Jowitt Whitwell**
b. 1858

♂ **Robert Jowitt Whitwell** – – – – – – –
b. 1860
m. 17 Apr 1884 Louisa Crommelin
 Brown (1859 -)
d. 1928, aged 68

body and mind'.[131] He had 'singularly white smiling teeth, and beautiful hands, which were equally neat and clever in tying up a parcel, posting a ledger, or making his own inimitably delicate pen-and-ink and pencil drawings'.[132] He applied his youthful energy, enthusiasm and considerable intellect to learning the family business. He had an excellent head for figures and not only kept meticulous accounts but also drew up detailed statistics of the local trade and information on business failures. Nonetheless, it is clear that his father did not place too heavy a burden on the young man's shoulders. During his fifth year in the business, John junior was still regarded as an apprentice, as he himself acknowledged during a visit to London in 1831:

> do very little business in town – 'tis rather tiresome, looking & looking & buying nothing, but this is my 'prenticeship.[133]

John junior may have been attending the wool sales, but this London visit seems have been taken up with sightseeing and visits to friends and relatives. His diary records trips to see members of the Marriage family and their flour mill near Cheltenham (a family concern which is still going), to Dorking and Box Hill, to the Armouries at the Tower of London and to the Zoological Gardens at Regent's Park, where he saw a Brahman bull, ostriches, an elephant, beavers and a lynx which was 'in a bad humour & scolded like a cat, only far worse.'

For some years, John Junior would ride or drive to the Huddersfield market every Tuesday to buy wool, leaving home at six or seven in the morning and returning at nine in the evening.[134] Visits to the London wool sales and to the company's far-flung customers could be an arduous and expensive business in the days before the railways. For instance, a trip to Kendal, Manchester and Stockport in 1832 would typically take about a week and cost about 3 guineas.[135] He would sometimes ride to Ilkley on a jackass. Even after the coming of the first passenger railway to Leeds in 1834, many places (including Kendal) could only be reached by road or canal and his diaries show that he was still travelling to London by coach rather than train in 1836, possibly because it was cheaper. He might also have regarded it as safer. A slow-moving train on which he was travelling was struck from behind by a new Galloway locomotive just outside Manchester in 1831. Although many people were thrown from their seats, no-one seems to have been seriously injured, but the consequences could have been very serious:

> If we had been quite standing instead of slowly moving, the consequences would have been dreadful – an engineer who was in the same carriage with me said we should have been smashed to pieces.[136]

As soon as John Junior was twenty-one in 1832, his father gave him £1000, made him a partner and changed the name of the company to

Fig. 14 Ilkley Bridge drawn by John Jowitt jr (undated).

Robt. Jowitt & Son. The initial terms of John and Robert's partnership were recorded by John in a memorandum:[137]

1832
9 mo. 15th.   Terms of Partnership: my Father & self,
     1st.   2/3 profit or loss belong to RJ.
          1/3 do—   do—   to J. J. Jr,
          after allowing Int. on their respective Capitals

     2nd.   That RJ take the risk of debts owing at
          the stocktaking 6 mo. 30th. 1832, and all monies
          accruing on previous bad debts, belong to RJ.

     3rd.   That the firm of RJ & Son pay £100 a year rent
          of the warehouse in Albion St. to RJ.

     (Signed by) RJ.
     1833. 5 mo. 17th   Jno: J. Jr.

Over the next few year, John junior applied himself hard to the family business and drew very little income: £50 in the year to the end of June

1833, although his share of the profit was £1695 6s 5d; £110 in 1834 when his share was down to £388 2s 8d; and £85 in 1835 when his share was £253 14s 11d.[138] These figures illustrate both the volatility of the wool trade and, because they left much of their capital invested, the partners' resilience in the face of a few poor years of trading. This was just as well, as there were many poor years.

John junior's drawings increased massively to £731 in the year ending 30 June 1836. There were two reasons for this. In the summer of 1835, he took a tour of France, Italy and Germany and in 1836 he got married.

LOVE, A GRAND TOUR AND MARRIAGE

John junior first met Deborah Benson, the eldest daughter of Robert and Dorothy Benson of Parkside, Kendal, in 1830 at the wedding of their mutual cousin, Maria Crewdson to John Eliot Howard.[139] When he visited his cousins the Dockrays in September 1831, they seemed to be pumping him for his opinion of her but, though he agreed that she was 'most beautiful, the flower of the Society,' both he and his cousin Rachel had reservations about her:

> ...praised DB's disposition, rather too hot tho', & romantic – says she wants stability – alas! what unstable things these young girls are.[140]

Not long after this, he met a particularly charming young Quakeress, Anna Elizabeth Dearman (aunt of his future son-in-law, John Dearman Birchall).[141] On 19 October 1831, he noted:

> Have just come home from a dinner party at cousin Jno. Jowitt's [John Jowitt senior's] – and such an intoxicating evening, talking all the time with Eliza Dearman & then walking home with her. She is a most superior young woman: – a delightful companion & so thoroughly affable, yet without the habit of condescending, which is so disgusting in some otherwise agreeable people – she is very intellectual – likes the people I like.[142]

It would seem that he was quite smitten with Elizabeth Dearman. Waiting in London for the 8 p.m. Edinburgh Mail one evening with Jacob Hagen,[143] John junior said, gave him

> the opportunity of pumping him about Eliza Dearman, & I think from what he and H[arry] C[hristy] say that he still has his eye on her, to say the

least of it. He was at Kendal some weeks ago & was during 4 days 3 times at Parkside & praised D[eborah] B[enson] famously...[144]

On 23 November 1831, he writes,

Saw E. Dearman off to London by the Union [stage coach]. She gave me a lingering look (I thought) & a hasty squeeze, & I felt or fancied, rather queer, after she was gone – pooh! – why fancy myself worse than I am – but I only journalize fancies...

One thing I am sure of, she is one of the most delightful women I know...[145]

While John junior was clearly attracted to Elizabeth Dearman, there was another woman who claimed his attention and was often mentioned in almost the same breath, Deborah Benson. As he recorded in his diary for the same day, it was not Elizabeth but Deborah that he dreamt of after seeing Elizabeth off on the Union stage coach. His feelings for Elizabeth had been immediate and powerful but those for Deborah had grown slowly and were more deeply seated. In a most revealing comment, John junior says of Elizabeth Dearman,

She is in mind and body everything one could wish; – if she had but a Kendal heart.[146]

Given that John junior's mother, Rachel Crewdson, came from Kendal and that four of his aunts married men with Kendal connections, it is hardly surprising that he should feel a fondness for that place and, by extension, for Deborah. A further, confusing, connection was that Deborah's brother George married Susannah Crewdson, daughter of Joseph Crewdson and John jr's great-aunt Rachael Jowitt, while Deborah's sister Rachel married Susannah's brother Robert. Rachel and Robert Crewdson lived in Wordsworth's cottage, Rydal Mount, for more than twenty-five years after the poet's death. As well as their shared link with the Crewdson family, it is possible that John and Deborah were distantly related through the Benson line. John junior's great-grandfather had married an Ann Benson, whose family came originally from Whicham, about twenty-five miles from Kendal.

49

Fig. 15 Rydal Mount, William Wordsworth's house at Ambleside, was home to Robert and Rachel Crewdson, cousins of the Jowitts, for more than twenty-five years. Drawing by John Jowitt jr, 30 June 1874.

By 1835, John junior and Deborah Benson were unofficially betrothed (both families seem to have been happy with the arrangement, but the formalities required by the Society of Friends could not be hurried) and were making plans to marry in late Spring 1836. John had wanted to marry in September 1835 but, as John noted,[147] Deborah was not in any hurry. Perhaps she should have been, because John junior seems to have had a roving eye, even if he was attracted by women who reminded him of her:

Sarah Binns – nice girl – good looking... Mary Salthouse... (sweet faced as ever I saw almost & a good bit like DB)...[148]

... I monopolized a very pretty irish girl, who charmed me with her likeness to my darling.[149]

nothing particular in the way of company except a smart good-looking girl who talked very fluent English, French, German, Dutch, & Italian, rattled away with the young men, & was hand & glove with Miss Story, the Badener, who proved at last, to be the Barmaid at the Pays Bas![150]

Before the wedding, there was another exciting event to look forward to; they were both to make a tour of Europe, not together, of course, but not entirely apart either. One hundred and forty years after his great-great-great-grandfather Richard Jowitt I died leaving his wife and children dependent on the charity of the Friends, John junior was embarking on a 'Grand Tour', the rite of passage of a young gentleman. It was a testament to the industry and plain-dealing of his father and grandfather that he could afford not only the expense but also the leisure to do so.

Deborah left for the Continent with her family in July, travelling to Ostend, Brussels, Ghent, Cologne, Waterloo, Geneva, Frankfurt, Mainz, Heidelberg, Chamonix, Lausanne and Milan. She had arranged to meet up with John junior in Berne, but he had plenty to do at home first, including finding a house. In any case, he probably needed a rest before setting out for Europe; he had arrived home on 20 June having travelled 560 miles around Britain in ten days.[151] It is also clear that the financial responsibilities that marriage would bring had been weighing heavily on his mind. On 2 July, he wrote,

> About 2 to-day, I balanced the Trade a/c, & with a trembling hand, & almost a failing heart, added up the two columns – I thought of D., – of my prospects, – of perhaps, several hundreds *loss,* & I did not really expect any profit at all.

> It is about £300, I leave blank, to correct any errors, which there may be.

> Oh! how very grateful ought I to be! I am in some degree. Lord forgive my ingratitude, & now enable me to *spend* aright that which thou hast given me.

He nearly had to cancel his Continental tour altogether when he was subpoenaed in a case relating to a will which he was believed to have witnessed, but it turned out that the witness was his cousin, John Jowitt senior. Events seemed to conspire to prevent his departure. On 31 August, he wrote,

> Something may *yet* put my going aside – Father hurt his leg at *my house* to-day, which might have done it, had it been a little worse, & I *fell over a dog,* which *might* have broken my leg; – in short everything teaches me that we live & breathe & move under a special providence.

John junior eventually escaped Leeds, arriving in Calais on 5 September. He stayed at the excellent Hotel Quillacq in the Rue Neuf, where Byron's friend Leigh Hunt had stayed in 1826, and dined on

> soup, fricaseed veal, ditto beef, fish, nameless stew, pastry, fruit, biscuits, & cheese, vin ordinaire, of the colour of claret, being our only drink.[152]

John junior was not impressed with Calais, which he said, had 'little to claim the traveller's notice, or repay him for a whole day lost there,' although he later regretted missing the church of Notre Dame. He was even less impressed with the gendarmes who

> looked miserable in their equipment & countenance, regular shilling-a-day-men. NB – they get seven sous![153]

He travelled from Calais by Lafitte's diligence (i.e. public coach), arriving at Paris's Barrier St Denis at 4 a.m. on September. John junior thought the approach to the city along the Rue St Denis 'by no means good: – perhaps equal to Shoreditch,' an area which he knew well from his business trips to London.

He took in all the usual tourist sights – Montmartre, the Napoleon Column, the Place Vendôme, the Louvre, the Champs Elysées, etc. – before attempting to book a coach to Berne. Lafitte's (which John supposed to be the largest coaching concern in the world) had no places for several days, but John junior presented a letter of introduction from John Capper, Secretary of the British and Foreign Temperance Society, to the Director, Captain d'Aublay. Surprisingly, perhaps, the letter carried some weight and Captain d'Aublay received him cordially and took him him to the Bureau d'Allemagne, where he successfully booked his journey at a cost of 75 francs and 39 centimes.

He could not get a bed at the Hôtel Meurice but found very comfortable lodgings at the Hôtel Windsor next door. Here they had English-speaking staff, although this does not seem to have mattered much as John junior's French was quite good (and he dots his diary, in a schoolboy way, with French words and phrases). Before leaving Paris on 7 September, he thought he caught sight of the King's carriage, remarking,

> He is not popular; – poor man! – king of such a people. They seem gay & light-hearted but *no religion*! How can such a nation prosper?

Leaving the capital via the Barrier Charenton, he remained unimpressed and regretted only that he had not received a letter from his beloved Deborah and had missed seeing the Bourse and the Père la Chaise cemetery, where, among others, that most unfortunate of lovers, Pierre Abélard, is buried. The road

> lay along the Seine for some time, on a wide paved road, flat, without hedges, & with rows of trees on either side, much resembling that from Boulogne to Calais, which is as wretchedly untempting as can be.

Rather inappropriately for a man whose very journey had been facilitated by the Temperance Society (headquartered at Aldine Chambers, Paternoster Row, London), he took this moment to remark,

> How much vin ordinaire has gone down my throat since I came to this merry land of France! Tell it not in the Aldine Chambers![154]

John junior's journey took him through Vesoul, Troyes and Bar-sur-Aube to the French border town of Belfort. An artist at heart, he remarked on the splendid mountain views and even the wooden and broad hats worn by the locals. After a journey of about 236 miles, he finally arrived at Belfort, where he supped on boiled milk which 'tasted right good English'. His German was good enough to allow him to chat easily with the hotel manager and waiter.

From Belfort, he travelled on the outside of the mail coach, a 'crazy vehicle,' with his feet in a wicker basket which did little to protect his feet from the driving rain. Entering Switzerland, he took heart from the idea that he was at least in the same country as Deborah. The journey through the 'curious old town' of Porentrieu and Delémont was very mountainous and for much of it he walked ahead of the coach. Arriving at Berne on the evening of of 9 September, he was disappointed to find that Deborah was not there, or even in Switzerland. A letter from her explained that they didn't expect to be in Berne until the sixteenth, having been detained in Milan for three days due to problems with their passports. He decided to try and surprise her party at Meiringen, arriving at Interlaken on the evening of 10 September, but they were not there, so he went back to Meiringen on 11 September. The next morning, his guide, Johann Schmidt, woke him at five so that they could go for a scenic walk through the snowy valleys, accompanied by a young Dr Ikins from Leeds. They travelled through the valleys of the Hasli to the Châlet at Handeck, which offered 'cheese, eggs, wine, potatoes, eau de vie, & all sorts of good things'. Here the medical man, whom John considered very intelligent, left their company to visit the nearby hospital. After lunch, John, guided by an eight-year-old Swiss girl from the Châlet, went to sketch the Handeck waterfall. Back at the Châlet, he met some travellers, three women and a man and he asked if they had seen Deborah's party. The women had not, but the man thought he might have seen them entering Altdorf, and John junior didn't know whether to go back to Interlaken, to stay in Meiringen or to head for Sarnen.

On Sunday 13 September, he visited the Reformed church where he remarked that the red kerchiefs and white sleeves 'had a pretty effect' but left because he could not understand the service, instead visiting the Reichenbach falls some fifty-six years before Sherlock Holmes fought with Professor Moriarty there. This time he was guided by a girl of six, 'a bright little thing'. He wondered if it was acceptable to sketch on the Sabbath but in the end the bad weather decided for him. He sent a messenger to Interlaken in search of Deborah's party, but the waiter woke him at 2 a.m. to report that there was no sign of them.

On the Monday, he decided to walk to Lungern but had only gone two miles when he was delighted to meet Deborah and her companions travelling from Lungern. They rode to Brientz and then crossed the lake to the Giessbach Falls before travelling on to Interlaken, staying at the Pension Hofstetter where they received a letter from John Wilson telling them that John's schoolfriend Wilson Lloyd had died of typhus, news that saddened Deborah and her father as much as John.

On Tuesday, John's twenty-fourth birthday, they set out again, passing the Jungfrau and Silverhorn. They rode the steep road up to the châlet on Wengen and they had hardly reached it when they saw an avalanche and heard its roar. John seems to have been oblivious to the deadly danger of these spectacular events:

> Three others fell while we were there, & one a very fine one. A cannon was fired & we had glorious echoes.

They then descended to the inn at Grindelwald. They were exhausted but had eyes and strength enough left to admire the splendid sunset, & then the rosy hues on the snow of the Wetterhorn – the Schreckhorn was in the background, beyond the second glacier.

The next morning, they rose early and left on horseback for the Grosse Scheidegg, stopping to admire the glacier, 'an amazing collection of ice mountains piled one upon another, many hundred feet high'. John and another of the party[155] crawled into a newly-formed cavern in the ice 250 feet high:

> this was the most curious place I ever was in. Thousands of millions of tons of ice above & around us, which might, for aught we knew, close as suddenly as they had formed the chasm we were in. The sight at the entrance was very beautiful. It was quite a Polar scene: the ladies each with an alpenstock, and all possible caution, walking on the field of ice, & we emerging from this fearful hole.

The passage of the Grosse Scheidegg itself was 'another glorious mountain ride'. There was much snow here and on the Wengen, and the descent was very steep, but they got down safely to Meiringen, seeing the glacier at Rosenlaui and the Reichenbach falls on the way. They stayed at the same inn, Le Sauvage, at which John had stayed before meeting Deborah and her party. The next day, they rode to Brientz for breakfast, and then on to Interlaken in the pouring rain. They devoted two days to mountaineering, 'as fine as heart could wish'. On Friday, John and Deborah managed to spend a little time alone together for almost the first time since they met up.

On Friday 18 September, they travelled by steamboat from Neuhaus to the Bellevue Inn in Thun, 'a beautiful & splendid hotel overlooking the lake'. On Saturday, they climbed the hill above Thun to admire the view before setting out in two carriages for Berne. John and Deborah had a carriage to themselves, for which the young lovers were very grateful.

Staying at the Couronne in Berne, 'a queen of cities' as John called it, they were so taken with the views from the tower of the cathedral that they ascended it a second time and

> watched the sun sink slowly behind the Jura, and all the landscape was wrapped in grey and purple long before the last rays of the sun had ceased to illuminate the mountains.

They went to an English service on the Sunday, given by Philipp Hirschfeld, a Jewish convert. Staying in while the others went to a French chapel in the afternoon, and then going for a walk together in the evening, John and Deborah were able to spend much of the day together. They were enchanted by the city's beauty:

> There are hundreds of acres laid out in public walks & pleasure grounds, on all sides of the city, entirely open & stocked with rare plants botanically named. We saw many of the inhabitants walking out, but not one shabby looking person, much less *one* drunk. From the ramparts & from these promenades we had a glorious view of the all-glorious snowy alps, which were again *all* perfectly clear, & the most perfectly beautiful sunset that could be conceived. We staid till it was 'all grey,' and the lamps were lit & the stars were shining when we reached the Couronne.

On Monday 21 September they arrived, after a journey of about forty miles, at Le Cigne in Lucerne. John thought the hotel splendid and the town old and feudal-looking. They watched the many poor pilgrims, men, women and children, travelling to and from the shrine of the Virgin Mary at Einsiedeln. On 23 September they travelled on, marvelling at the scenery on the way:

> The first view was as overpowering as the brightest imagination could have painted. 300 miles of mountains (in circumference) burst at once on the eye, some say 500; about half the glorious panorama is snowy, the other half, which embraces the Lakes of Lucerne, Zug, Sempach, & two other small ones is comparatively flat, & richly wooded; – bound on the north by the Jura & Vosges, & east by the Sea[156] of Constance.

They next descended the Righi mountain to the villiage of Goldau, which had been buried under a landslide in 1806. They were following in the footsteps of Wordsworth, who had come this way fifteen years before,[157] but it was some verses by Felicia Hemans not the Lakeland poet which John junior quoted (if imperfectly remembered):

> I looked on the mountains, a vapour lay,
> Shrouding their tops in its dark array –
> But thou lookest forth, & the mist became
> A shield & a mantle of living flame!

They again encountered the Einsiedeln pilgrims and, with his Quaker heritage, John was offended by the 'Crucifixes and & shocking paintings ... a sad contrast with the scene around.'

John and Deborah did not have much time alone, but he says,

My dearest is most kind, & tho' our select opportunities are scarce, we enjoy much of each other's society. How privileged I am – I often think of the dear ones at home, & wish I could know how Father is.

His concerns were soon allayed when he found a long letter from home waiting from him in Zurich on 24 September. His father's arthritic arm was 'much the same' and his sisters sent contributions for the holiday. In Zurich they stayed at what John judged to be a bad inn called Le Corveau, and called upon the famously artistic Gessner family.

Next they travelled on to the dramatic Rhein falls at Schaffhausen, between Neuhausen am Rheinfall and Laufen-Uhwiese. They quite equalled John's expectations, 'which had previously been lowered by common report':

There is a great body of water, but the height is only 40 feet! – yet we get so near it at the foot, that the effect is very grand.

It's spoilt by a cockneyfied sort of castle, & seats, & galleries, & a board, worse than all, inscribed, 'Sonnez pour voir la chute du Rhin,' & when you have sounded, & seen a creature that opens you the gate, you must pocket out a franc each person. The boatmen too are extortionate – indeed travellers seem to be considered only as fair game for all sorts of imposition.

In the evening they arrived in Bondorf, Baden, having said a sorrowful farewell to Switzerland. They next travelled through the Höllental, a valley which John described as 'beautiful even after Switzerland,' to arrive in Freiburg. John considered the cathedral's stained glass the finest he had ever seen, but, just as he disapproved of the 'shocking paintings' of the Einsiedeln pilgrims, he considered the altar to be 'disfigured by a picture by Holbein, of the crowning of the virgin by our Saviour, the Father, & the Holy Ghost.'

From Freiburg they travelled on, via Kenzingen (where they breakfasted) to the 'pretty town' of Offenburg. From here they could see Strasbourg Cathedral, the Vosges and the Black Forest. After another pleasant day's ride, they found themselves in the 'handsome little metropolis of Karlsruhe' (then the capital of Baden):

> It is the neatest & the cleanest town I ever saw. All the buildings are white with stucco, or cement & there are many Grecian fronts & porticos. Soldiers at every turn. The Duke's palace is a splendid pile, & behind are noble gardens, open to the public, where we walked by moon light. In front, groves of orange-trees then a large square, then streets diverging like rays, & others crossing these at right angles. The town is nearly surrounded by a dense forest, & the approaches are lined with rows of Lombardy poplars in some directions. [In] spite of my dislike of this tree, these roads are handsome. The gateways are very fine. The Duke is *very* popular, & every thing reminds you that you are under his Eye...

On 1 October, they travelled to Mannheim, stopping on the way at Schwezingen and Heidelberg. John was most impressed with the ducal residence in Schwezingen, in the grounds of which

> Bronze dragons and sea-monsters throw up noble fountains, & groves of orange trees, & long avenues of acacias & mimosas with sweet little vistas thro' them fill up the enchanting landscape.

> My darling & I hurried over them, & left the party in the Mosque of Mecca (an exact model, ⅙ the size, of extreme beauty) & took a voiture to Heidelburg while they leisurely surveyed the gardens...

Six miles of straight road brought them to Heidelberg, of which John wrote,

> ... the town is nothing, the Inns are nothing, the people are nothing, the castle is everything – We lost no time, in climbing up to it, by a steep flight of steps, – thro' several broken walls, & vaults, & soon gained the esplanade where a fine view of the country bursts upon you. The Geissberg & the Königstuhl however rise so close, as to intercept the view except towards the west & south. The course of the Neckar may be traced nearly to Man[n]heim, & the Rhine is just visible in 2 or 3 places. The Vosges bound the horizon, a fine, long chain...

John and Deborah were so enchanted with Heidelberg that they dallied there until after dark, taking a coffee at the Prince Carl Inn before

Fig. 16 Heidelberg Castle drawn by John jowitt junior when he and Deborah
returned there in 1875.

hiring a carriage to take them to Mannheim for three florins, where Deborah's party awaited them. John described the palace there as,

a frightfully ugly building, dull & heavy – red stucco, & whitewash disfigure the walls; the theatre was burnt down some time ago, & is still a ruin, & every other part's fast going to decay. The apartment of the dowager duchess of Baden, Stéphanie [de] Beauharnais (Joséphine's daughter) are splendid. The gardens too, & grounds, are very fine.

From Mannheim, they travelled by steamboat to Mainz and were disappointed by the scenery on the way, which John called 'very dull & flat'. Of Mainz he wrote,

every other person you meet is a soldier. There are 6000 Austrians, & as many Prussians, & the town belongs to Hesse Darmstadt. The streets are fine, the buildings neat, but nearly all barracks.

They then headed down the Rhine to Cologne and on to Nijmegen, enjoying the view, although John regretted that,

The Rhine has disappointed me, but Switzerland is to blame for that. All tourists should see the Rhine *first*. From Rheinstein, the King of Prussia's castle, to Bonn, is most of the way, fine scenerey, but the vineyards spoil the hills, & they are all small & stinted, Nothing however can exceed the beauty & endless variety of castled craggs, the Drachenfels, the Rolendseck, the Katz, the Maus, Schönberg, & Rheinfels, & Bingen, & St. Gaor, & Oberwessel – with a hundred others, all storied with legends & romance, in all stages of decay & of all kinds of architecture, & in all sorts of situations. Every turn displays some new beauty, & the steam boat goes so fast that you have hardly time to admire a castle before it is gone.

In Cologne, John visited the perfume business founded by Johann (or Giovanni) Maria Farina, vendor of the real and original *eau de Cologne*. The journey on to Nijmegen ('a neat little fortified town, with the finest cabbages to sell I ever saw') was dull, with nothing to interest them all day, so that they spent most of the time reading in the cabin and distributing a few tracts. Whether they were religious, anti-slavery or temperance, John does not say, but one can only imagine that they were not enthusiastically received by the more frivolous of their travelling companions. Among these were some members of Baden society, the nineteenth-century equivalent of 'gay young things':

> 3 young persons, just returning, fresh from the vortex. Poor, poor fashionables! Miserable their slavery!

More to his liking was General Sir E. Carroll, a Badener who knew his Bible, and also, from his time in Malta, the Anglican missionaries William Jowett and Joseph Wolff. The General's two sons had studied at Philipp Emanuel von Fellenberg's pioneering school at Hofwyl in Switzerland and John thought they were a credit to the school, though he fretted that they were to join the army forthwith.

By 5 October they were in Rotterdam, whence they sailed for England aboard the paddlesteamer *Batavier*. The headwind and choppy seas made John very queasy. Nonetheless, he was able to get a good night's sleep and he rose the following morning to find they were off Margate in a thick fog. Arriving in London at 3 p.m., they took a boat to the Tower of London and then walked to Broad Street where a letter was waiting for John assuring him of his father's good health. There was sad news, too: their friend Joseph Gurney had lost his wife some days before, his cousin Jane Crewdson had died from typhus, and Wilson Lloyd, another cousin, was also dead. They stayed with the Howards in Lordship Lane, Tottenham.

John arrived home on 10 October to find his family well, although his father had suffered a serious fall which had rendered him unconscious for half an hour, and he was still troubled with headaches. Another friend was dead – Jonas Hobson had met an accident a fortnight ago while riding home drunk. John refrained from moralising, perhaps remembering his own liberal consumption of *vin ordinaire*.

The anticlimax started to dawn on John on Sunday 11 October not helped by his father delivering two striking sermons on preparation for death, and, of course, he missed his darling Deborah. He started looking at furniture in Kendell's for what was to be their new home at 10 Blenheim Terrace, fretting, 'this will be an expensive job – wish it was well paid for, & the *choosing* over.' Indeed, he seems to have become rather obsessed with furniture, staying up late with John Howard to discuss it, when he was in London in November. At the same time, he could not stop thinking about Deborah, writing on 14 November,

Fig. 17 Robert Benson of Parkside (1780–1857), Deborah's father.

This day week, – nay, in *4 days,* I hope to be with her in body as well as in mind. May God almighty bless us both in, & for, & to each other!

Sadly, it was not to be, as the Priestmans were staying at Parkside and there was no room for John. He would have to wait another fortnight to see his love, arriving at Parkside on 1 December, delighted to find Deborah alone. He was in a very poetic state of mind during this visit:

I am here once more, & tho' the 'chill December' has robbed the trees of their 'leafy honour,' the scene is lovely still. It is midnight, & the bright full-orbed moon is reigning, in peerless majesty over a sky of the richest

Fig. 18 Parkside, Kendal, drawn by John Jowitt junior in 1849.

Fig. 19 Parkside, Kendal, drawn by John Jowitt junior in 1868 after it had been rebuilt in a whimsical gothic style.

hue, & all is still & hushed save the gentle murmur of the pebbly brook which flows into the pond.

John was both excited by the prospect of marriage to Deborah, provisionally planned for April 1836, and slightly concerned. On 3 December he wrote,

> At Meeting this morning almost overwhelmed with kind words from kind friends – afternoon and evening alone, – too much trifling during the former, yet much sweet intercourse, & talk about our future plans – Dearest Dora! if only she could see exactly as I do about the 4th. mo. I sometimes fear that she is not happy in the prospect, because so desirous to retard it.

On 13 January 1836, he received the keys to his new house, and by 16 January he had bought most of the furniture, exclaiming, 'May we find God in this house!' but he was back at Parkside by 20 January. While there, he and Deborah went with a neighbour, Edward Wakefield, to look at the nearby Sizergh Castle (once the home of Henry VIII's sixth wife, Catherine Parr),

> which looks habitable, a very dark wainscotted, haunted-chambered, tapestried, turreted castle.

Remarkably, it was not until 15 February 1836 that he received his credentials from the Kendal Meeting ('the official first step!') and it was not until 26 February that he could write, 'I am declared! – the intended husband of Deborah Benson!' They had undertaken a European grand tour together before their engagement, with chaperoning so relaxed that they could spend many hours alone together. It may have been unorthodox, but it was a very good way to make sure the young couple could form an enduring relationship.

Between visiting Huddersfield for the sales and furnishing the marital home, John found time to gather 8,000 signatures for an anti-slavery petition. He also found time to read the salacious details of the Mellin v. Taylor criminal conversation case at York:

> They have acquitted a married man, who committed adultery, in open day. What will things come to?

All the time, he was counting off the weeks until they would be married. Even then, there was a formal betrothal at the adjourned Monthly Meeting in Kendal on 14 April before the actual wedding. By Saturday 29 April, the house at Blenheim Terrace had been furnished and heated, so that John could entertain his family to dinner.

John and his mother travelled to Kendal on 2 May and his father and sisters followed on 4 May. They dined that evening at Sizergh Castle.

The following day, John wrote,

I am married! God give me grace to keep my marriage vows in everything! – This is, I believe, my first desire.

We have been been greatly helped, most abundantly favo[u]red in meetings – thro' the day, & this evening – a delightful religious opportunity. 52 to tea.

Dear Elisha Bates preached an excellent sermon on the resurrection of our Lord, & the Doctrines in connection with that glorious event; then shewing the application of these truths to the things of this life, & particularly the changes which take place in the 'dear domestic circle,' – shortly after he sat down, we rose, & my dearest D. (*mine* now, by emphasis) said most beautifully: – we have universal Commendation...

John's father recorded that about 20 people dined at the Benson family's house, Parkside, (twenty-one according to John) and fifty-two took tea there:

After the latter repast a religious opportunity took place in which much excellent counsel was handed forth by E. Bates and H[annah] C[hapman] Backhouse,[158] mainly to the new married pair, setting forth in a striking manner the danger of their becoming lukewarm in their love and allegiance to their Saviour, of allowing other things to occupy that first place which He ought to have in their hearts and affections.[159]

The presence of Elisha Bates has an interesting aspect. This leading Quaker minister and printer from Mount Pleasant, Ohio, was disowned by the Society of Friends after being baptised in 1837 at the height of the Beaconite controversy (see p. 67). He later became a Methodist minister. One leading American Quaker described Kendal as 'the hotbed of Beaconism'.[160]

Compared with the European tour of the previous year, John and Deborah's honeymoon was a modest affair, but equally energetic. On 6 May, they left at about 7 a.m. in 'a snug post-chaise,' arriving in Low Wood in time for a second breakfast. From there they walked to the bowling green by Helvellyn. Then they travelled in an open carriage, rounding the head of Lake Windermere and turning up Langdale ('the pikes most magnificently Swiss towering above us') before ascending Loughrigg Fell, a ride of four or five miles to Grasmere, which John considered 'a charming little lake'. They rested briefly at the Wythburn Inn at the base of Helvellyn before riding on to Keswick where they booked accommodation, and then took a boat on Derwentwater to visit Barrow Bay and Lodore Falls, 'which looked but a miserable driblet.' They then walked about a mile up Borrodale before returning to Keswick in time for tea.

The following day, they travelled towards Patterdale in a 'huge open caravan of a coach', enjoying the views of Skiddaw, Saddleback, the Helvellyn Fells and the vale of St John's before spying Ullswater and through Gowbarrow Park, past Lyulph's Tower to Aira Force, 'a beautiful little fall, hono[u]red by a rainbow or two in the spray'. There John made a hasty sketch of Ullswater. Not satisfied with these exertions, the newly-weds climbed the steep western flank of Place Fell after tea, observing the large slate quarries there.

On 8 May, their first Sunday as a married couple, they attended the little church at Patterdale, which bore the date 1631.[161] The sermon was pedantic and, perhaps in consequence, the congregation meagre. On the Monday, they came over Kirkstone to Ambleside, admiring Sockgill Force on the way. Just before reaching Ambleside they were lucky to escape injury when a wheel came off their coach, and they had to complete the journey on foot.

On 10 May, only five days after the wedding, they were back at Parkside, warmly greeted by their families.

After the honeymoon, they settled down at 10 Blenheim Terrace, Leeds, two or three minutes' walk from his father's house on Carlton Hill.[162] The Jowitts and the Bensons were already close, but no-one could then have guessed what an important role Deborah's family and particularly her youngest brother, William Thomas Benson, were to play in the Jowitt family's fortunes.

To move from the beauties of Parkside in Kendal to a major urban area like Leeds must have been daunting, even for a well-travelled young woman like Deborah. Indeed, John had asked himself on their honey-moon,

> How can I ever make up to my own sweetest darling, for such friends, &
> such a home! But she is so kindly disposed to look on the bright side of
> everything, that I have hopes.

And it seems that his hopes were not unfounded. Deborah was imme-diately taken with her surroundings, revelling in a 'delicious' eight-mile walk to Meanwood (the brook not yet polluted by the Jowitt fellmon-gery), and back through Woodside and Headingley, declaring 'I had no idea of so much beauty in the near neighbourhood of Leeds.'[163]

THE BEACONITE CONTROVERSY

When people belong to a persecuted minority, as the Quakers had in the seventeenth and early eighteenth century, the persecution can create a sense of solidarity. When that persecution ends, so may the solidarity, as the Quakers were to find out, not once but twice in the nineteenth century. In 1828, a rift at the yearly meeting of the Society of Friends in Philadelphia had created upheavals among the Quakers which were to have an important, if indirect, effect upon the Jowitt family. A char-ismatic preacher called Elias Hicks had for some years been propagat-ing a doctrine which was unacceptable to some of the more traditional members of the Society. He argued that the 'Inner Light' (as the Quakers call the conscience) rather than the Scriptures should guide one's be-haviour. To some this was Quaker orthodoxy, to others it was heresy. He also taught that sinful behaviour resulted from God-given human nature, not from the influence of Satan. His followers, the Hicksites, had been banned from the yearly meetings, but in 1828 they turned up anyway and the ensuing confrontation caused a schism amongst American Quakers.

In England, many Quakers, influenced by the growing evangelical movement, already believed that the Friends should re-assert the Scrip-tural basis of their faith. To them, it seemed that, by continuing to em-phasise the Inner Light, Quakers were in danger of embracing Hicksian heresy. Quakers had traditionally disapproved of water baptism and their old minute books are full of derisive references to 'sprinkling', but many of the new generation were worried that this objection to baptism was at odds with Biblical teaching. John Wilbur, a very conservative Quaker minister from Rhode Island visited England in 1831 and was deeply concerned by the growing evangelism within the British Friends. John junior, who heard Wilbur preach, described him as 'an agreeable minister.'[164]

In 1835, John junior's uncle, Isaac Crewdson, a well-loved preacher in the Manchester Meetings, published a pamphlet entitled *A Beacon to the Society of Friends,* in which he emphasised the paramount authority of the Scriptures and warned of the 'desolating heresy' that had 'lately swept thousands after thousands of our small section of the Christian Church into the gulph of Hicksism and Deism.'[165] John Junior immediately bought a copy of the *Beacon,*[166] but he was far more than a casual reader. On 3 March 1835, he noted in his diary that he had sent Isaac suggestions for the second edition and had received a very kind reply, and on 14 March he had almost completed preparing the index for it.

Isaac Crewdson followed up the *Beacon* in 1837 with *Water Baptism: an Ordinance of Christ,* a tract which argued that baptism was an essential element of Christian faith. John junior seems to have already accepted the argument for baptism. When Edward Wakefield of Kendal was 'sprinkled' by a Presbyterian minister in 1836, John wrote in his diary,

> May he be a better Christian! He has written a book, on the subject, collating the scriptural evidences. I think it is decisive on the point...

One American Quaker, in a letter to the 'agreeable' John Wilbur wrote of Isaac Crewdson and Joseph John Gurney (another leading English Quaker and abolitionist),

> Each of these writers has published doctrines essentially at variance with those of the religious Society of Friends, which being carried out and adopted, must unavoidably undermine Quakerism; and it is self-evident, that whosoever *openly defends or advocates either of these men,* commits himself to and identifies himself with the doctrines of the man whom he thus defends against the Society, and the more especially so, if that defence be made upon the occasion of others opposing such doctrines. This course has been unhappily pursued, and to a fearful extent, by prominent members of New England Yearly Meeting, to whom allusion will be hereafter made.[167]

While he may have been tactless, it seems clear that Isaac Crewdson only intended to nudge the Friends towards a more Scriptural, less Hicksian, theology. Ironically, he precipitated a schism in British Quakerism as profound as the Hicksian controversy in America. According to John junior, Isaac had sought a 'quiet imperceptible evangelizing of the Society'. To Beaconites, as Crewdson's followers were known, there seemed to be a 'bad spirit' among their opponents at the Quarterly Meeting in April 1835, although 'Uncle I[saac] C[rewdson] was enabled to keep his temper admirably'.[168]

While at Parkside on 18 April 1835, John junior recorded in his diary,

Terrible accounts from Manchester Q[uarterly] M[eeting] a most painful time; party feeling at its height: a red-hot conservative Candidate appointed: – alas! alas! this is in a "Society of Friends"! Let us blush for the name we bear, thenceforth call ourselves *Quakers*... Father will be distressed.[169]

And, indeed, John junior's father, Robert, was distressed, even more so when it became clear that John junior and his sisters were inclined to follow Isaac Crewdson's lead. John junior wrote,

Weeping. Have just had a most painful conversation with Father, – Mother was by, & took my part occasionally. Oh it is dreadful! Lord God, for Jesus sake, open thou his eyes, & *ours*, to see the *very truth*. Mother seems much distressed, & Father all but overwhelmed; – & when I tried to assuage his grief, he said it was prophesised &c – about the division of families – 'Mother-in-law against her daughter-in-law &c'...[170]

Robert Jowitt published *Thoughts on Water Baptism* in 1837, a non-confrontational but rather well-reasoned rebuttal of Crewdson's arguments on the matter. The last two paragraphs of his tract sum up his argument:

How full is the testimony of John the Baptist to the spiritual character of the gospel dispensation – 'And now the axe is laid unto the root of the trees: therefore every tree which bringeth not forth good fruit, is hewn down and cast into the fire. – I indeed baptize you with water unto repentance; but He that cometh after me is mightier than I, whose shoes I am not worthy to bear: He shall *baptize* you with the Holy Ghost and with fire, whose fan is in his hand, and He will thoroughly purge his floor, and gather his wheat into the garner, but He will burn up the chaff with unquenchable fire.' -Matt. iii. 10, 11, 12.

Here is a baptism essential to us all – the baptism with the Holy Ghost and with fire. – Whatever may be our views as to water-baptism, let us ever bear in remembrance the emphatic language of the apostle, 'Not by works of righteousness which we have done, but of his mercy He saved us, by the washing of regeneration and renewing of the Holy Ghost.'[171]

Despite efforts on all sides to heal the rift, including the work of a Committee of Elders, one of whom was John's old schoolmaster, Josiah Forster, controversy raged and families split along doctrinal lines. Other Friends left the Society for more practical reasons, such as John junior's friend John Hattersley who was, said John junior as early as December 1831, 'no longer one of us, in principle at least.'[172] Hattersley's objections to Quaker doctrine seemed weak and superficial to John junior but he yearned to attend Oxford University, which, as a non-Anglican, he was precluded from doing by the Test Act. Hattersley never did go to Oxford but he broke with the Society anyway.

Some of the Beaconites felt obliged to leave the Friends before they were forced out. One such was Robert Rathbone Benson, a cousin of Robert Benson of Parkside, who resigned at the Liverpool Meeting on 25 October 1836.[173] Others, like John junior, his sisters Elizabeth, Susanna and Rachael, and his wife Deborah, were shown more compassion. A minute of the Brighouse Monthly Meeting of 17 August 1838 records their separation from the Friends (which had actually occurred in 1837). It is headed 'Disownments for Baptism at Leeds' and states:

For these dear friends, individually, we, nevertheless continue to feel a very tender regard; and desiring that grace, mercy, and peace may be with them, we affectionately bid them farewell in the Lord.[174]

As one contemporary writer remarked, it was 'very forebearingly ex-pressed, and is a model in this respect for all excommunicators.'[175]

John junior then became a staunch supporter of the Congregational church in South Parade. His brother Robert Crewdson Jowitt left the Friends in 1843 and his sister Esther left in 1850.[176] John's parents re-mained in the Society of Friends throughout their lives but, despite their differences, the two generations stayed on the warmest terms. Robert and John still worked together and, while they no longer worshipped together, they maintained compatible social values. Father and son were both renowned for their charity and good works, especially in the fields of temperance and slavery. Many other Quaker families, such as the Lloyds and the Bensons of Parkside, split along similar generational lines.

JOHN JUNIOR, FAMILY MAN

John Jowitt junior and Deborah had six sons, all but one of whom, Robert Benson Jowitt (usually called simply Benson), died within hours of birth[177] and five daughters, Susan Maria, Rachel Elizabeth (Lily or Lillie), Anna Dora (Dolly), Emily (Em) and Florence (Flossie), the last two born at 5 Beech Grove Terrace, after their move there in 1850. The new house was only just around the corner from Blenheim Terrace, within what is now the Leeds University campus.

It is clear from the memorials to John junior left by Susan Maria, Anna Dora and Benson that it was a very close and loving family.[178] Although John was very busy, he always had time for his wife and children, taking them on outings to Kendal, Ilkley Moor, Bolton Woods, Whitby or Redcar. Susan Maria remembered being driven by her father to Kendal in her grandfather's carriage.[179]

John junior's great warmth and sense of fun are revealed in his letters, even those written near the end of his life:

> Just a line – just a globule, a nutshell of love, a grain of common sense, or what other smallest thing can be compressed to this tiny sheet.

> When I was young we had elephant paper for long letters (and sometimes crossed them all over besides). Imagine a mouse paper for the love letters of mice, and all intermediate vertebrates accommodated by half-quarter sheets of all kinds of sizes...[180]

He took great trouble over the children's education, sending RBJ to boarding school and employing a live-in governess, Martha Jackson, for the girls.[181] He would rise early in order to give Latin lessons, not only to Susan Maria, but also to the governess. Incapable of holding a tune himself, he paid for all his children to have music lessons. The quality of the girls' education (as well as her innate ability) is reflected in the fact that Emily passed the Cambridge Examination for Women, the equivalent of a university degree, when she was only twenty, gaining first class honours.[182] In 1859 the family moved again, to the 750-acre Harehills Grove estate in Potternewton, between Chapeltown and Chapel Allerton. Most

Fig. 20 Harehills, John junior's home. Reproduced by kind permission of Leeds Library & Information Services.

of the land was sold for development but thirty acres of parkland were retained around the handsome neo-classical house which had been built for the woollen merchant James Brown in about 1817.[183] On leaving Leeds in 1900, R. B. Jowitt sold the house and grounds, then referred to as the Harehills Park Estate, to Leeds City Council for £36,000.[184] The house (which is now known as Potternewton Park House) is still standing.

John's eldest daughter, Susan Maria, married Theodore Howard at Harehills on 26 April 1860, the wedding costing her father the considerable sum of £97 10s 3d.[185] Lily married Theodore Crewdson on 17 August 1864. On 4 May of the following year, Anna Dora married David Howard, brother of Susan Maria's husband Theodore. David and Anna Dora's daughter, Helen Elizabeth Howard, married the barrister Sir Charles Stafford Crossman. One of their children was Richard Crossman. Like his fellow Labour minister and distant cousin, William Allen Jowitt (see Appendix XIII), Crossman went to New College, Oxford.

Robert Benson Jowitt married Caroline McCulloch in the Anglican St Nicholas's Church, Durham on 6 September 1865. The wedding was officiated by the Reverend Arthur Horsley Hughes, the bride's brother-in-law.[186] She was the daughter of Samuel McCulloch MRCS, who had died suddenly from a lung infection in 1853 when his daughter was just

Fig. 21 Samuel McCulloch MRCS (1793–1853).

nine. Emily's husband, Dearman Birchall, was later to describe Caroline as 'jolly and has fine eyes but rather gross, heavy looking and very stout'.[187] Samuel McCulloch had attended medical lectures at the University of Edinburgh before studying surgery in London under the eminent surgeon and anatomist, Sir Charles Bell FRS (of Bell's Palsy fame).[188] In some ways, Caroline's and RBJ's fathers could hardly have been more different. Samuel had been commissioned as a second assistant surgeon in the Royal Ordinance Medical Department and had served with the Royal Horse Artillery during the Peninsular War, at the battles of St Sebastian, Nives, Nivelle, Orthes and Toulouse. He was also stationed at Bordeaux briefly before being transferred to Canada in 1814. Here he treated the wounded of the battle of Queenston Heights and other engagements of the so-called War of 1812 between Britain and the United States. He was placed on half pay on 27 January 1817 and then practised in Liverpool for many years, both in private practice and as surgeon to the Liverpool Workhouse and Fever Hospital. Having served under the Duke of Wellington in Spain, McCulloch enjoyed recounting stories of him and the gory battles in which he was engaged.[189] By contrast, John Jowitt was a

73

lifelong pacifist, and when he saw the Duke of Wellington in 1852, he wrote,

> surely a humble Christian widow... is a far worthier object of envy than this octogenarian Captain, who has drunk his fill of glory here, and may have (for I would not judge him) or may perhaps *not* have a better inheritance above.[190]

Caroline's father had been a Baptist, so it would seem that both she and RBJ had gravitated from a non-conformist background to the established church. It would have been a far less dramatic change, though, than John junior's break with the Friends in 1837.

Emily, like RBJ, converted to Anglicanism, to accommodate her spouse's adopted faith. She married John Dearman Birchall, nephew of the seductive Elizabeth, on 22 January 1873. He had previously married Clara Brook, daughter of William Leigh Brook of Meltham Hall in 1861, leaving the Quakers and joining the Church of England in order to do so. Sadly, Clara died in 1863, leaving him a daughter, Clara Sophia. Dearman bought Bowden Hall near Gloucester in 1868, becoming the local squire. Although there is no suggestion that it influenced his proposal to Emily, Dearman was very impressed with the rumours of John junior's great wealth. He recorded in his diary on 8 June 1870 that Jowitt was 'said to be worth many hundreds of thousands of pounds'.[191] Dearman, who was extremely wealthy himself, was the son of Samuel Jowitt Birchall who had been a wool stapler in the Jowitt & Birchall partnership. His father had died at Springfield House, Leeds, on 8 January 1854 leaving £35,000.[192] He and Emily were, of course, distant cousins, his grandmother, Anna Jowitt, being Emily's great-great-aunt. The Jowitts of Leeds, despite their growing wealth and John junior's Grand Tour, had tended to eschew notions of gentility (Grace, Joseph Jowitt's widow, being the first to style herself a gentlewoman) and continued to be actively involved in business. In contrast, Dearman had largely divorced himself from the day-to-day concerns of commerce (although not in any way from the income derived from it)[193] and he had embraced the manners and customs of an English country gentleman.

John junior's youngest daughter, Florence, didn't tie the conjugal knot until 1889, after her father's death, when she married Arthur Paine Baines, son of Edward Baines junior, the campaigning editor of the *Leeds*

Mercury. Even if she had married sooner, there would have been no danger that Harehills would fall silent; the first of John's twenty-seven grand-children was born in 1862[194] and the park was to ring with the sounds of children and grand-children for many years. Furthermore, RBJ and Caroline settled at Elmhurst, very close to Harehills Grove.

ROBERT RETIRES

John junior's father, Robert, had always felt that his business concerns should be subordinated to his religious duty, although his business life was always intermingled with religion and the simple pleasures of travel, whether sightseeing or visiting friends. Robert's attitudes on the balance between commerce and religion were reflected in those of his fellow Quaker, William Forster, in a letter to Robert of 2 May 1805:

> 1805. 5th mo. 2nd.—It is a nice matter, or at least so it has appeared to me, to follow business so that it shall neither occupy too much nor too little of our attention. To pursue it with too much indifference is the way to contract idle habits, which not only injure the mind but enervate the body. On the other hand, to follow it with avidity, as though our whole comfort, both in this life and that which will ensue, depended on the greatness of our temporal acquisitions, leads to many evils. It seems to fetter the mind to the earth, contracts our ideas as well as our hearts, and I believe tends greatly to impede the most important duty of this life, that of a preparation for another.[195]

Robert had long promised himself that, if at all possible, he would retire from the firm at the age of sixty. As he gradually handed the running of the business over to John junior, he reduced his capital invested, from over £20,000 in 1835 to just over £11,000 in 1842 and by 1845 John junior's share of the profits increased from one third to a half to reflect the changed workload.[196] He completed his retirement as planned at the age of sixty in 1844.[197] It should not be imagined, however, that Robert lived a hermetic or unworldly life in retirement. His letterbook for 1853–6 is filled with endless share dealings.[198]

Fig. 22 Robert Crewdson Jowitt by an unknown artist.

ROBT. JOWITT & SONS

The 1830s and 1840s were lean years for the partnership (see Appendix IV) and John junior confided his concern to his diary in 1836:

> Dear Robert came home on the 5th day [Thursday] with Father from the Q[uarterly] M[eeting]. I enjoy the thought of having him with us, & I hope we may get trade enough for two, tho' I fear *work* will be scarse enough.[199]

Nonetheless, the company officially became Robt. Jowitt & Sons (plural) when RCJ became a partner in 1842,[200] at the age of twenty-one, although many of the ledgers and correspondence only start referring to Robt. Jowitt & Sons in 1844. Given the financial situation, it is not surprising that there is little evidence of his active involvement in the firm before that time. Indeed, there is no occupation listed for him in the 1841 census when he was twenty, a stark contrast to his brother who had joined the company before his fifteenth birthday. He became more active in the company in 1845 when he made trips for the company to London,[201] following that with journeys to Liverpool, Germany and London in 1846.[202] Sadly, he died at his father's residence, Carlton House, on 3 October 1847 at the age of 26 after six days of 'intestinal inflammation'.[203] This was not during one of Leeds's occasional outbreaks of cholera and there may be a possibility that he contracted something from the London docks, for he had only returned from the metropolis on about 24 September.[204] As would be imagined, his death was much mourned by John junior and the rest of the family.[205] A brief death notice but no obituary was published.

During one of the (not infrequent) periods of uncertainty and economic turbulence in the year to the end of June 1837, the company recorded a trading loss (i.e. excluding the considerable sums they earned in interest from their customers, but which were probably swallowed up in interest charges paid to the bank) of £6,242 10s 9d, more than wiping out the previous three years' profits. A more prolonged period, between 1839 and 1849, was extremely difficult for the English wool trade and the economy and general, in part due to problems in the United States, where a series of banking crises and a catastrophic fall in the cotton price caused a steep fall in demand for many imported products. Despite this, Robert Jowitt & Son performed quite well during this period and saw no losses on the scale of those that they had suffered in 1836–7.[206] John junior's drawings from the company grew from £296 3s 4d in 1839 to £614 12s 3d in 1847. However, it is notable that he paid many business expenses, such as travelling to sales, from his personal account. In 1836, he even spent £2 15s 8d (about £240 at current values) of his own money on tea, butter and candles for the office.[207] In later years, the intermingling of his and his son's personal and business accounts became increasingly complicated.

John junior carefully recorded details of business failures in the wool trade. In 1839 alone, he listed 48 failures, in 1840, 103 (twelve of which were wool staplers), slowly recovering to a mere six in 1848. Some of them were of vast proportions and could have a domino effect. When the worsted spinners, Hindes & Derham of Leeds and Dolphinholme (Lancashire), collapsed with liabilities of £225,000, it brought down six other companies[208] and even led to a tightening of discounting rules for Bank of England branches.[209] The Jowitts' main bankers, Becketts, lost more than £20,000 to bad debts during this period. It is a reflection of Robt. Jowitt & Sons' great caution and their habit of maintaining close business relationships with their customers that their losses from bad debts were so small – the worst year being 1843 with bad debts of about £600. However, their cousins' firm, John Jowitt & Co. lost £1,000 in 1846.[210]

Given the continuing uncertainties in the wool trade, it is unsurprising that John junior, like his father before him, sought to diversify, often into railways. He held shares in the Alva Railway, Border Union, Wear Valley Railway, Leeds and Thirsk Railway and many others. He also owned stock in the slate quarry at Festiniog, and the Leeds Oriental Bath Company. Further afield he invested in the East India Navigation and Canal Company and a number of American and Canadian railways.

John junior, who acted as administrator in a number of business insolvencies, was one of the many businessmen who later opposed the Bankruptcy and Insolvency Bill, arguing in his capacity as Vice President of Leeds Chamber of Commerce that 'the principle of levying a tax for the benefit of the general community on those who have already suffered the loss of their property is highly objectionable.'[211] Another campaign by the Chamber of Commerce which had John junior's whole-hearted support was the formation of the Post Office Savings Bank. Although it wasn't the first to propose it, the Chamber put its considerable weight behind the scheme on 7 November 1860 and the Chairman and John junior were selected to be its delegates at a meeting with the Attorney General.[212] The savings bank came into effect on 16 September 1861.

When, in 1865, the Leeds Chamber of Commerce proposed allowing workers in the wool warehouses to finish work at two on Saturdays, John junior's temperance notions prompted a rather less enlightened attitude, and he warned that 'the effect of having more time might be to increase immorality.'[213] Despite John junior's opposition, the Chamber passed the motion.

Like his father, John junior was very active in good causes. He worked in the soup kitchens in times of hardship, and for many years he would make weekly visits to the poor on behalf of the Benevolent Society. Both John junior and Robert gave generously in response to disasters at home and abroad, such as the Indian famine of 1860–61.[214] Also like his father, he gave generously to Leeds General Infirmary. He was one of the founders of Ilkley Bath Charity Hospital, one of several hydropathic establishments in that town. He was on the committee of both Cookridge Convalescent Hospital and Leeds Reformatory for Boys at Adel.[215] Contemporary reports seem to indicate that the Reformatory achieved considerable success in rehabilitating the young delinquents in its care.[216]

Father and son campaigned tirelessly against slavery and were so well-known in this regard that a confidence trickster calling himself Curtis

Buck, who gave lectures in Huddersfield and Gildersome on his supposed experiences of slavery, made a point of claiming to be well-known to them before intimating that 'a little pecuniary aid would not be unacceptable.'[217]

After he left the Society of Friends in 1837, John junior and his friend John Wade founded the Leeds Town Mission at Blenheim Terrace. One of the Town Missionaries would breakfast with them every week. Later, they were frequent visitors to John junior's magnificent house, Harehills, at Potternewton. John junior was secretary of the Mission for more than 43 years, and then President until his death. He was also involved in the Religious Tract Society and the Bible Society.[218] He was also a strong opponent of the opium trade with China, although his motives and those of his fellow campaigners were not quite those one might expect. At a meeting held in December 1858, John junior suggested that the opium trade was damaging the export of woollen cloth in China while R. N. Fowler of the Church Evangelization Society argued that it was impeding the spread of Christianity in China and India.[219] John junior shared their taste for evangelising in foreign lands and supported missionary activities in India[220] and among Irish Catholics.[221] Although John junior was a mild and modest man, the zealotry expressed by those around him was uncompromisingly arrogant. The Mayor of Leeds, at a meeting of 1 February 1858, attended by John junior, Edward Baines, the Rev. Walter Farquhar Hook and others, suggested that the Indian Mutiny had resulted from withholding the blessings of Christianity from the population, and called Hinduism 'one of the greatest abominations on the earth'. Speaking of Hindus and Muslims, the Rev. Hook said it was 'their principle to propagate their religion by fraud or by force'.[222]

John junior served for a while on Leeds Town Council and was on the School Board (as Vice Chairman 1870–7 and Chairman 1877–9) but he had no real liking for politics and declined the opportunity to stand for parliament in about 1870.[223]

The disruption of trade resulting from American Civil War caused some hardship to the West Riding wool trade and much greater hardship to the Lancashire cotton workers. In what was then a remarkable example of friendship between the wool and cotton trades and between the commercial interests of Yorkshire and Lancashire, John junior, together with William Beckett, the Reverend Edward Jackson and John Whiting, formed the Leeds branch of the Cotton Districts Relief Fund and he acted as one of its secretaries between 1862 and 1865. The generosity of the local population (who were themselves suffering from a downturn in the wool trade) was such that £8,000 was pledged on the day the Leeds branch was created, including an initial £150 (or £25 per month for 6 months) from Robt. Jowitt & Sons.[224] By 25 November, more than £16,000 had been pledged.[225] John junior took it upon himself to visit Preston. Drawing a sharp distinction between the deserving and undeserving poor, his attitudes displayed the familiar blend of compassion and prejudice which characterised liberal Victorian opinion:

> Avoiding the low or Irish quarters, and wishing rather to see some of the houses of the more respectable working classes, I visited alone, and quite at haphazard, eight or ten houses in succession, in Brunswick-street. In every house I found distress, arising from the times, and in every one relief, more or less complete, was being given.[226]

RETIREMENT AND DEATH

John junior had, quite naturally, hoped to retire by the time he reached his early sixties, but the large amount of his capital tied up in the business and the arrangements he had made in his will meant that he felt obliged to continue for longer than he would have liked.[227] Apart from falling ill in 1869 from which he rapidly recovered, he had always enjoyed good health and was able to carry on for some years, gradually withdrawing from the business from 1875. He sold the warehouse to Robert Benson Jowitt in 1878 for £6,000.[228]

In 1879 John junior's wife Deborah suffered a serious bout of illness and, though she recovered, the shock affected his own health and he suffered a stroke in 1880, the year in which his daughter Lily died; even then, he recovered and returned to work, not finally retiring until 1 January 1884, at the age of 72.[229] Only three weeks later, his daughter Emily died, aged only 31.

For the last few years of his life, his mind lost its former sharpness and his body became more frail, until the only thing which he seemed to retain of his former self was his piety. He fell seriously ill on 15 December 1888, lying unconscious for a fortnight before dying on the morning of Sunday 30 December. The *Leeds Mercury* described him on the day of his funeral as 'a man whom to know was to revere'.[230]

The memorial at St John's Church, Roundhay, reads:

Fig. 23 The Jowitt family at Carlton House, Leeds, in about 1858:
1 Robert; 2 his son John junior; 3 John junior's wife Deborah; 4 Robert's daughter Elizabeth; 5 Robert's daughter Susanna; 6 Robert's daughter Esther Maria; 7 John junior's daughter Susan Maria; 8 John junior's son Robert Benson; 9 John junior's daughter Rachel Elizabeth; 10 John junior's daughter Ana Dora.

In loving memory of John JOWITT
born September 15th 1811
died December 30th 1888.
'I shall be satisfied when I awake with thy likeness'
Also of Deborah his wife
born September 10th 1813
died August 8th 1893.
'I will bless the Lord at all times'

Robert Benson Jowitt (1839–1914)

Benson's come home again,
Welcome our brother then,
   Benson's come home!
Sing we all merrily,
Heartily, cheerily,
Glad we are verily,
   Benson's come home!

John Jowitt jr[231]

Robert Benson Jowitt studied at University College London, where he came first in the junior collegiate prize for English in 1856.[232] For some reason, he never completed the degree course and appears to have gone into the family firm later that year.

Although the company had been importing increasing quantities of wool from the continent and the colonies, home-produced wool still accounted for half its business up to 1873.[233] It also started importing mohair for the worsted trade from South Africa in 1880, something which was to become an important part of the company's business. By the time RBJ's father died in 1888, the company dealt almost exclusively in wool from the empire. The exact dates at which the company founded its various overseas branches are not shown in the surviving records. In most cases, the company first entered into partnerships with local companies. In 1868 Robt. Jowitt & Sons bought land at Hindmarsh, Adelaide (South Australia) and by the 1870s it was doing business in Australia with: Younghusband & Co. and Lyell & Gowan in Melbourne; David Guthrie & Co. in Geelong (Vic), Ballarat (Vic) and Kingston (SA); Thomas Guthrie in Rich Avon East and West (Vic), Brim (Vic), Keilira (SA) and Mount Muirhead (SA); and the Hon. Saul Samuel in Sydney. By 1885 it had its own branch in Melbourne. In South Africa, its earliest dealings seem to have been with Lamb Brothers of Port Elizabeth in 1880. Its first South African branch was in operation by 1888. In New Zealand, the company was trading with Johnston & Co, of Wellington by 1877 but did not have its own branch (at Dunedin) until 1891.

The shift to overseas sourcing was a general one in the industry, but it is still remarkable that the company under RBJ's leadership was able to entirely reinvent itself so successfully.

Fig.24 Robert Benson Jowitt (1839–1914) and his wife Caroline (1844–1921).

Robt. Jowitt & Sons was described in 1888 thus:

Among the largest and most renowned houses engaged in this trade, there are few that have such a valuable reputation as that of Messrs. Robert Jowitt & Son, [sic] a firm that not only is justly entitled to be considered a high representative house, but is certainly one of the oldest in the trade... The scope of the firm's operations is mainly directed to the trade in Australian wools, and is exceedingly large and comprehensive. The firm have an office in London, in order to facilitate the despatch of business and to carry out every detail of their transactions upon a very extensive scale. Numerous travellers represent the firm in all parts of the United Kingdom, especially in the North of England, where the business connection is very extensive and widespread.[234]

As well as a shift in sourcing, another, equally important change took place in the company's business, the move away from wool dealing to taking wool on consignment. In the eighteenth century and early nineteenth century it had been a largely logistical operation, sourcing wool in various places, arranging transport and selling it elsewhere. This had always been a speculative business, and the large fluctuations in profits

84

Fig. 25 The fellmongery at Geelong, Victoria, 1889. The company's name appears on the right-hand building. This is the earliest known picture of a Jowitt overseas subsidiary.

reflected this, but with the coming of railways and reliable international shipping services, the logistical element became less important because, in theory, anyone could do it, and the business became almost entirely speculative and very, very risky. However close wool staplers were to their customers, they could not always anticipate sudden changes in trade due to natural disasters or man-made events, such as the American Civil War and the Franco-Prussian War (the outbreak of which caused an immediate doubling of the wool price, and its conclusion with France's defeat, an equally sudden collapse). Sourcing from distant lands increased the time taken to bring the wool to market, and thus the risk. Taking goods on consignment allowed Robt. Jowitt and Sons to reduce its risk and at the same time subsidise their overseas buying operations. Unfortunately, it also reduced the potential profits. Consignments, initially from Australia only, rose rapidly from 1875. Consignments from South Africa started in 1878, and by 1879 more than half of the wool shipped by Jowitts was on consignment. However, by 1881 they had started reversing the trend with more direct purchasing.[235] The apparent advantages of taking goods on consignment were largely illusory. Wool brokers were most eager to consign goods when the market was saturated, where an outright sale at a good price was difficult. They would expect Jowitts to pay them an advance based on their own, often over-optimistic, valuation. One such consignor was David Guthrie & Co. which owed Robt. Jowitt & Sons nearly £9,000 in 1876.[236] Thus Robt. Jowitt & Sons would still be risking large sums of money in falling markets.

The company experienced recurrent problems in both Australia and South Africa with their agents overbuying or paying too much for wool. Some of this may be blamed on poor communications, although the final stage of the telegraph cable linking Britain to Australia was completed in 1872, allowing messages to be sent in seven hours rather than several months.[237] The problem, at least as perceived from Leeds, was that the overseas staff, including managers, thought they knew better than Head Office. On 31 March 1892, RBJ expressed his frustration very plainly to John Thomas Gibson, the Manager of the Melbourne Office, a man who seems generally to have been held in high regard both within Jowitts and within the wider wool industry:

> I must tell you plainly how bitterly disappointed we all are to find that again this year, owing to no fault of our own, we are in for another heavy loss by Melbourne buying for selves – And I cannot understand how you could so much disregard our instructions as to buy some 1300 bls. for selves, in addition of course to Barwon products, unless you thought 'The home people are in a panic. I will act on my own responsibility' – You also in your letter that Mr. Tate replied to last week say 'low as your limits have been they were not low enough' – as if now that things had turned out badly you wished to [thrust?] the responsibility on us. Ever since last Sept. we have all been convinced (& none more so than Mr. Fred & Mr. Tate) that prices would inevitably fall from sale to sale till a much lower basis was reached. Exactly what we foresaw has taken place. We wanted you to buy nothing for our selves except at v. low rates.

Oct 16. wire limited you to 20% for merino & 15% for xbred below last season's rates. Has this been strictly followed out?[238]

RBJ was even less satisfied with one South African employee, writing to James Lawford, father of the Cape Town manager, on 6 June 1894,

For some months I have been uneasy about the so-called branch at Cape Town, feeling since it was no good to us, & fearing that it would be no good to your son & it was only my strong wish to make this final trial a sufficiently long & thoroughly fair one, that prevented my interfering earlier to stop it. The last mail however brings such strong opinions from Port Elizabeth on the subject[?] – and these fall in so exactly with what we have all felt at home – that I have decided that we must at once give up that business. I am therefore writing both to Cape Town & P. E. to that effect. As we have no opening of any kind for your son we are obliged, most reluctantly, to terminate his engagement with us & my letter to him tomorrow will convey that intimation... I don't want to make matters more painful to you by giving you fully all my reasons for this step. I have nothing but good will towards your son, who I feel sure, has always tried to do his best, & whose character is above suspicion. He has not however shown any of the special capability needed, & he has repeatedly acted contrary to positive instructions both from P. E. & from home. We have been paying over £200 a year for a branch we don't want, & which has only lost us money, simply because of my previous instructions to our people to find a berth for him.[239]

Nonetheless, RBJ offered to pay the young man three months' salary and his passage home, as long as he travelled within two months of receiving his notice. This was extended to four months in a letter to James Lawford on 8 June.[240]

The continuing difficult conditions in the wool trade are reflected in the extremely large and thorough list of trade references kept by Jowitts during the period from 1881 to 1900.[241] This massive tome, as large and heavy as a lectern Bible, contains references on some fifteen hundred firms, about twenty percent of which failed during the two decades covered. References were provided largely by other members of the trade and by Bradford Old Bank and the Schimmelpfeng Institute. The Bank's references were often very guarded and, rather in the manner of the credit rating agencies before the credit crunch, it described as 'respectable' many companies which subsequently failed.

Like his father, Robert Benson Jowitt was a very energetic and able businessman. Unlike his father, he had not been brought up a Quaker, and this manifested itself in a couple of ways. His father seems not only to have preached temperance but to have been almost teetotal (after his youthful encounters with *vin ordinaire*), but RBJ bought fine clarets, whisky and Tetley's beers.[242] The last-named could have been for workmen, but the whisky and claret were certainly for his own consumption. He also showed less inclination towards proselytising, although he was active in the Anglican church. While Benson did not engage in quite as wide a wide range of charitable and evangelical activities as his father had, concentrating particularly on the Leeds General Infirmary which his father and grandfather had helped support for many years, he certainly contributed greatly to the local community. Given the size and complexity of the business which he built up, and the vagaries of the wool market, it is remarkable that he could find time for any outside commitments. He served on the board of the Infirmary for twenty-eight years becoming both Chairman and Treasurer of the Weekly Board in 1882. His resignation, due to leaving Leeds for retirement in Tunbridge Wells,[243] was called 'perhaps one of the greatest losses the General Infirmary suffered before 1900 commenced'. A resolution passed by the fellow members of the board stated:

> The Board of the Leeds General Infirmary desire to place on record their deep regret at Mr R. B. Jowitt's resignation, and to express their high appreciation of the services which he has rendered for twenty-eight years as a member of the Board, and of the conspicuous ability with which he has filled the office as chairman and treasurer for nearly twenty years. During these years many important changes in the constitution of the infirmary have taken place; large additions have been made to the buildings and the number of patients treated, and of the staff, have been greatly increased. The board are sensible that to Mr. Jowitt's tact and influence it is largely due that these changes have been successfully accomplished, and that his exertions, aided, as they have been, by his personal generosity, and by the confidence of the public in his management, have maintained the Institution in a sound financial position; and that his helpful sympathy for the cause of medical education, closely identified with the interests of the Infirmary, had a great share in securing for the Leeds School of Medicine the important position which it now holds in the medical world...[244]

Fig. 26 The portrait of Robert Benson Jowitt paid for by public subscription on his retirement from the Chairmanship of the Board of Leeds General Infirmary in 1900.

After his retirement, the inhabitants of Leeds subscribed to a fund to place a portrait of him in the boardroom and to establish the Jowitt Pension Fund for nurses and other workers in the Infirmary.[245]

Robert Benson Jowitt was also a justice of the peace for both the City of Leeds and the Skyrack division of the West Riding (as his cousin, Edward Jowitt of Eltofts, had been), and was unanimously chosen as Chairman of the Bench in 1893. In that capacity, he introduced the Scottish form of oath in which a hand was placed upon the Bible, in preference to the English tradition of kissing it, which he considered unhygienic.[246]

He was Chairman of the Leeds District Nurses Institution and was on the committee of the Cookridge Convalescent Hospital. As a member of the council of the Yorkshire College, the forerunner of the University of Leeds, he helped to oversee its amalgamation with the Leeds School of Medicine. Another of his medical interests was the Royal Albert Asylum for Lunatics and Imbeciles, the only such institution in the country at the time.[247]

One charitable involvement which saw a certain amount of opprobrium heaped upon RBJ's head was his role as Honorary Secretary of the Leeds Ragged School Society and Boot Blacks Brigade, a charity chaired by the then Prime Minister, Lord Palmerston, which offered free education to the poor. A Mr W. S. Thorne had proposed giving a benefit for the Ragged School at the Princess's Theatre, Leeds, and had apparently advertised the event without first consulting the proposed beneficiaries. This drew a polite letter from RBJ, saying that the character of the institution forbade them from accepting any share of the proceeds from a theatrical production.[248] To some, this smacked of puritanism. One commentator remarked:

> It is said that men resemble their grandfathers far more than they suspect,—the habits of their bodies rather than their souls making the chief difference. It strikes us, however, that large numbers of them resemble rather their grandmothers. Certainly it is undeniable that gentlemen daily thrust themselves before the public with such striking old-womanly qualities and habits of thought, that it is difficult to account for their escape from the petticoats and pattens suitable for such dear, good, silly souls.

> We do not wish to say anything offensive of Mr. Jowitt, the hon. secretary of the Leeds Ragged School. We take it for granted, that in writing his letter to Mr Thorne he acted for others as much as for himself; and therefore

Fig. 27 William Thomas Benson (1824–1885).

we shall not say anything of a harsh nature to Mr. Jowitt personally. But the supposition which we give him the benefit of, forces upon us the unpleasant conclusion that we must have in Leeds—and we say this notwithstanding the good work these men are engaged in—some of the silliest, most stupid, and impertinent specimens of piety to be found in connection with any system of religion on the face of the earth. It is intolerable that these people, by the puerility and disgusting offensiveness of their acts should bring a sublime religion into contempt,—a religion which, through being bewildered by pious conceit, they can neither feel no understand...

If these modern Pharisees were not drunk with spiritual pride, as partially educated Englishmen, they would know that the greatest human intellect the world holds knowledge of was exercised in creating the best specimens of the finest dramatic literature the world possesses. Who can stand up and say that the faculty given by God to Shakspeare had not its tendencies from the same divine source!... Intellectually, there is as great a difference between Shakspeare and the pious insects of Leeds, as between Micromegas, the giant of Sirius, and the little men who, when he came to this earth, crawled upon his thumb nail...[249]

And so the attack went on for many paragraphs. It must have been extremely hurtful to the 'pious insects,' men of the church or of business like RBJ and his father, rather than members of the university-educated elite. In reality, though, it was a well-educated group. As we have seen, RBJ had excelled at English, and John junior had benefitted from a first-class education at Tottenham. Among the other members of the Committee were Edward Baines M.P., who had attended New College, Manchester, the Reverend G. W. Conder, author of *Free Press versus Free Speech,* and James Kitson (later Lord Airedale) who studied Chemistry at University College London.

RBJ's wife Caroline was also active in charitable matters, being the Treasurer of the Leeds Ladies' Association for the Care of Friendless Girls, 'a small and unobtrusive but most valuable society' which offered training, guidance and housing to girls who might otherwise be tempted into 'crime and vice'.[250]

THE BENSON CONNECTION

Although the family no longer belonged to the close-knit Quaker community, they still enjoyed a very close relationship with a number of other ancient Quaker families; many of these had left the Friends at about the same time as the Jowitts, and probably most of them (like the Bensons, Braithwaites, Rathbones and Birchalls) were related to them by marriage. Among the most important of these was the extended Benson family, whose family banking business was eventually to merge with Kleinwort, Sons & Co. to form Kleinwort Benson in 1961. The banking members of the family were mostly quite far removed from Robert Benson Jowitt's grandfather, Robert Benson of Parkside,[251] but the connection was important nonetheless, and sometimes led the Jowitts into ever more exotic and potentially risky investments. Commercial relations between the Jowitts and the Bensons existed from at least 1816 when Cropper, Benson (a partnership founded by James Cropper and Robert Benson's brother Thomas) acted for John Jowitt & Son in relation to wool imported from Philadelphia.[252] Later both the Bensons and the Jowitts bought American railway shares and bonds through their mutual cousin Isaac Braithwaite of the stockbrokers, Foster & Braithwaite.

One of Robert Benson Jowitt's most important contacts with the Benson family was his mother, Deborah's, younger brother William Thomas

Benson. This enterprising young man left Kendal for Manchester where he teamed up with a Scot, William Blythe, who had studied Chemistry in Glasgow under Professor Thomas Thomson FRS.[253] In partnership, they formed Blythe and Benson, a company producing chemicals for the textile industry, and with an incongruous sideline as dry-salters. They also seem to have been directly involved in calico printing. The company was a very substantial manufacturer of bulk chemicals, often both hazardous and noxious. Their original factory, in Church near Accrington, caused so much pollution, in that age of smoke and grime and in an area not renowned in the Victorian era for the freshness of its air, that the partners were fined for the nuisance in 1849.[254] When their offices and warehouse at New Cannon Street, Manchester, were damaged by fire in 1850, it was suggested that the 'certain powerful acids which stood in carboys on the floor' might have been the source.[255] Despite these problems, the company's Church factory was advanced enough to be cited in an 1857 chemistry journal as an example of the modern method of producing sulphuric acid and soda,[256] of which it produced several thousand tons per annum. Their other products included garancine, a dye created by treating madder root with sulphuric acid, used at their new works at Holland Bank, Oswaldtwistle.[257]

When his father, Robert Benson of Parkside, died in 1857, William sold up his share in Blythe and Benson and moved to Canada with his wife, Helen (née Wilson). In Montreal, William met Thomas Aspden, an Englishman, who drew his attention to the lack of a starch factory in Canada. William and Thomas went into partnership and set up business in Edwardsburg, Canada West (now Cardinal, Ontario), a small town on the St Lawrence which had been settled by loyalists from New York, Pennsylvania and New Jersey in the wake of the American War of Independence.

More than a century after the founding of the starch company, W. T. Benson's great-grandson Mr F. T. B. Jowitt was phoned by an official representing Cardinal Town Council. It transpired that W. T. Benson had left his land to the council for a public park and, being a cautious man (especially in financial matters), had inserted a clause into the agreement saying that the land would revert to the family if it were put to any other use. The town wished to use fifteen acres of the land for expansion but could only do so by buying it from the family. Mr Jowitt's visions of great wealth from the sale of fifteen acres of prime, inner-city land were quickly dashed by the official who gave him to believe that Cardinal, far from being a thriving metropolis was something of a one-horse town. Never having been there, he had to accept the official's word and the land exchanged hands for C$50,000, of which his share was the princely sum of C$345.[258]

There was no corn (i.e. maize) grown in the Edwardsburg area and locals must have thought the two men mad when they built a corn-milling factory there, but the location was ideal. William realised that there was a plentiful supply of corn in the United States and that the Great Lakes, St Lawrence river and the newly-constructed Grand Trunk Railway would all help transport the raw materials and finished products. Water power was supplied by Lock 26 of the Gallops Canal.[259] The main products of the mill were starch for food and clothes, both in bulk for the trade and for domestic use, and glucose. No doubt his experience of providing bulk chemicals for the textile industry in Britain stood William in good stead. When Aspden sold up in 1860 and returned to his native Lancashire, Benson & Aspden became W. T. Benson & Company and, in 1865, the Edwardsburg Starch Company. Later still it became Canada Starch, with the trading name Casco. It is now a subsidiary of Corn Products International Inc. and mostly produces food products. The name W. T. Benson & Company was retained for the Benson family's import/export business in Montreal, which, amongst other things, imported wool from Robt. Jowitt & Sons.[260]

William Thomas Benson had proved himself a very successful and influential businessman, becoming a member of the Canadian Parliament in 1882. Not unnaturally, Robert Benson Jowitt and his father seem to have regarded William highly and to have drawn upon his investment advice and his connections with the wider Benson family. On 30 October 1883, for instance, RBJ wrote to William saying that his father would like to follow up his suggestion to sell some mortgages and give the money to William to invest with a return of not less than 6%.[261]

William also seems to have put his nephew in touch with the other Benson financial interests in North America, notably Constantine W. Benson, and with Close Brothers, the London merchant bank, who were involved together in the Iowa Land Company, investing in the fertile land made accessible by the Chicago, Iowa and Nebraska Railroad Company as they built westward from Illinois in the 1850s. In fact, so many British people invested in the land that it caused concern among Americans that they were being recolonised. The Jowitts bought 1,750 acres from Close

Brothers, Sibley, Osceola Co., Iowa in 1884.[262] However, as soon as July of that year, RBJ was writing to say that he would like to sell the lot 'as I have a demand for further capital for use here,' although he added that if he could be sure of selling all in two or three years, he would be happy to hold on to the land.[263] Nonetheless, Robert Benson Jowitt still had a square mile (640 acres) of Iowa in 1889 which he sold, using the proceeds to buy, through C. W. Benson, 1,760 acres of virgin land in Minnesota at $6.50 per acre. He also owned land in Riverside, California,[264] a city about sixty miles east of Los Angeles on the Santa Anna river founded in the early 1870s. It was the site of North America's first successful citrus plantations. This may have been one of John junior's investments as the temperance ethos of the city would have appealed to him more than to Robert.

W. T. Benson died suddenly in 1885, leaving a son, George Frederick, and daughter, Helen Dorothea. George went on to head both Canada Starch and W. T. Benson & Company, while Helen was to play an important part in the future of Robt. Jowitt & Sons.

OTHER OUTSIDE INVESTMENTS

As well as the investments made through the Bensons, Robert Benson Jowitt owned shares in the Alford Estate Co. of New Zealand. The company had been formed by a Mr Grant to buy the estate from Mr Grant himself, a fact which, in itself, might have given investors pause for thought. When the company lost money for investors, Grant promised them that it would pay 5% dividends for ninety years, failing which he would make up the shortfall himself.[265]

RBJ also had shares or bonds in the the Montana Gold and Silver Mine, Mundy Droog Gold Mine, Indian Consolidated Gold Mines, the West Australian Land Company, Buenos Aires Great Consolidated Railway, the Uruguay Pastoral Association (which was wound up in 1890)[266] and the London and Argentina Bank.

Nearer to home, his investments included the Yorkshire House to House Electricity Co. Ltd (founded 1894 and bought by Leeds Corporation in 1897, its history being described as 'amazing efficiency by a private company and amusing incompetence by the city council'),[267] the Leeds Girls' High School, the Colosseum Company Ltd (Leeds) and the South Market, Leeds. However, domestic businesses could be just as risky as overseas ones and he recorded a loss on a cosy-sounding Leeds company called Cocoa Houses Ltd.[268]

Some of RBJ's most interesting and most disastrous investments were those in G. R. Portway & Co. and G. Portway and Sons. They demonstrate both his wish to diversify within the wool trade and the complex and ever-shifting disposition of liabilities between various members of the family and the partnership. G. Portway and Sons was a company which controlled the rights to a number of patents granted to George Portway's younger brother Herbert, a Bradford worsted cloth manufacturer and loom-maker, for mechanical and electric stop machines, intended to stop looms whenever a warp thread broke. Herbert was a prolific inventor and a founder member of the Blackpool Tower Company Ltd.[269] One of his devices featured in the 15 January 1878 supplements of the *Textile Manufacturer* journal.[270] The patents were bought on 23 January 1878 by Robert Benson Jowitt, G. R. Portway and and John Frederick Wilson[271] (who married Portway's sister Charlotte in 1865 – like RBJ, he was a great-grandson of John Jowitt II)[272] for £10,000, of which RBJ paid £6,450, Portway £2,250 and Wilson £1,300. The patents were potentially very useful and very valuable and Grant & Hudson agreed to pay a royalty of 15s on every one of these machines they made.[273] It was arranged that when the purchase money was paid off, the partners would receive royalties, presumably in proportion to their investment. A Mr P. Hunter was employed as an agent in Manchester at £200 p.a. and J. Bradbury at a salary of £150 p.a.[274] Further money was needed and RBJ lent the partnership £1,500 in November 1878 at 5% interest, and a further £1,400 through Portway. John junior agreed to stand one third of the loss up to £2,000. The venture did not fare well and John junior paid RBJ £2,000 in March 1879, ending his liability. However, he gave RBJ a further £1,000 in December 1880, and RBJ wiped off £4,000 of debts, 'giving up the thing as virtually useless,' although he stayed in the partnership until 1883. Even then he allowed his name to be used for a further three or four years 'until the capital was paid off".[275]

RBJ became the sole partner in the other Portway business, G. R. Portway & Co., on 1 October 1878, agreeing to pay Portway a salary of £700 and William Musgrave Wood[276] a salary of £600. The company consisted of a substantial worsted and woollen mill (Swinnow Mill, Bramley), a warehouse, and a cloth merchants in York Place, Leeds. This seems to have been the Jowitts' first involvement in cloth-making since the eighteenth century.

An inventory in the possession of Mr F. T. B. Jowitt shows Swinnow Mill to have consisted of: a wash house; boiler house and tenter room; top tenter room; willey house (where the wool was teased and oiled); two scribbling houses; 'Attic No. 5'; horizontal engine room; 'firing place'; large engine house; weaving shed containing thirty-two power looms and several warping machines; cloth mill with milling, washing and wringing machines; perching room (where cloth was stretched); three weft warehouses; a shed full of miscellaneous machinery, including 'sundry old iron', two tons of railway sleepers, a weighing machine, a two-wheeled dog cart and four-wheeled Phaeton; fud (woollen waste) warehouse; 'No. 2 Mungo Shed' (since there is no No. 1, it seems likely that they drew no distinction between fud and mungo) with two more tons of railway sleepers; pattern weaving shed complete with high-pressure steam engine and pattern looms; two store rooms containing everything from spare drive belts to old wine; stable complete with bay mare (Kitty) in foal, and a dark bay of fourteen hands; 'Fud Place near Henhouse'; yard with several wagons; field with drying poles or racks, two pairs of stocks 'out of use without tappits', an old cow shed and pig stye, and sundry old cast iron; two offices, one of them designated private; and a warehouse at Stanningley containing two redundant power looms. The whole was valued at £5,223 0s 2d on 3 October 1879.

Capital for the purchase of about £17,000 was lent by Robt. Jowitt & Sons, but at RBJ's personal risk. On 1 October 1879, Portway joined RBJ as an equal partner in the mercantile arm of the company only, Portway putting in £3,000 and Robert £6,000, but on 31 December 1890 he dissolved the partnership in the cloth merchants and Portway became sole proprietor of this operation.[277] RBJ wiped out Wood's debts and paid off bonds on the warehouse and mill to the value of £8,100, of which £3,000 had come from his father. He then reorganised the business without the warehouse but continuing to run the mill, employing Wood at a salary of £365. The warehouse, which by then was equally owned by Portway and RBJ, was leased to G. R. Portway & Co. and Wilson.[278]

RBJ evidently gave up Swinnow Mill as unprofitable after a few years. It was run for some time by William Townend and Joseph Kelsall, trading as William Townend & Co., Worsted Coating and Cloth Manufacturers, until 1897 when Kelsall took full control.[279]

♂ **Robert Benson Jowitt**
b. 1839 at Leeds
m. 06 Sep 1865 Caroline McCulloch
(1844 - 1921)
d. 1914 at Tunbridge Wells, aged 75

♂ **John Herbert Jowitt**
b. 16 Jul 1866 at Elmhurst, Leeds
m. 16 Feb 1892 Rinah Mary Jane Hales
(1869 - 1949)
d. 14 Mar 1908 at Wellington, aged 41

♂ **Frederick Robert Benson Jowitt** – –
b. 1892 at Dunedin
m. 14 Sep 1921 Kathleen Margaret
Lupton (1895 - 1949)
d. 29 Sep 1965, aged 73

♂ **Eric Benson Jowitt**
b. Nov 1894 at Dunedin
d. 16 Feb 1895, aged 0

♀ **Dorothy Rinah Benson Jowitt**
b. 1896 at Dunedin
+. Reginald Taylor (1873 - 1959)
d. 1971, aged 75

♀ **Mary Caroline Benson Jowitt**
b. 1901 at Carlton Colville
d. 25 Oct 1987 at Newcastle upon Tyne,
aged 86

♂ **Frederick McCulloch Jowitt**
b. 1869 at Elmhurst, Leeds
m. 22 Aug 1900 Helen Dorothea Benson
(1867 - 1952)
d. 19 Sep 1921 at Hollins Hall, near
Harrogate, aged 52

♂ **William Thomas Benson Jowitt** – – –
b. 01 Nov 1901 at Leeds
m. 1929 Margaret Jean Law (1906 -)
d. 1941, aged 40

♂ **Robert Benson Jowitt II** – – – – – –
b. 02 Nov 1901 at Leeds
m. 1929 Audrey Haverfield Stanton
(1904 -)
d. 1966 at Bradford, aged 65

♂ **Richard McCulloch Benson Jowitt**
b. 1905 at Leeds
d. 1970, aged 65

♂ **Robert Jowitt**
b. 1870 at Elmhurst, Leeds
m. 1895 Adèle May Simpson (1868 -)
d. 05 Dec 1945 at Winchester, aged 75

♂ **Robert Lionel Palgrave Jowitt** – – –
b. 1899 at Leeds
m. 1930 Dorothy Marion Hartley (1905 -
)

♂ **Edward Maurice Jowitt**
b. 26 Sep 1874 at Elmhurst, Leeds
m. 12 Sep 1899 Edith Simpson (1878 -
)
d. 1954 at Bridport, aged 80

♂ **Anthony Thomas McCulloch Jowitt**
b. 1900 at Leeds
m. 1930 Doris Anderson (1896 -)
d. 14 Sep 1977 at Great Barrington,
Massachusetts, aged 77

♂ **John Alan Jowitt** – – – – – – – – –
b. 1904 at Leeds
+. Dawn Marsh (1922 - 2007)
d. 1996, aged 92

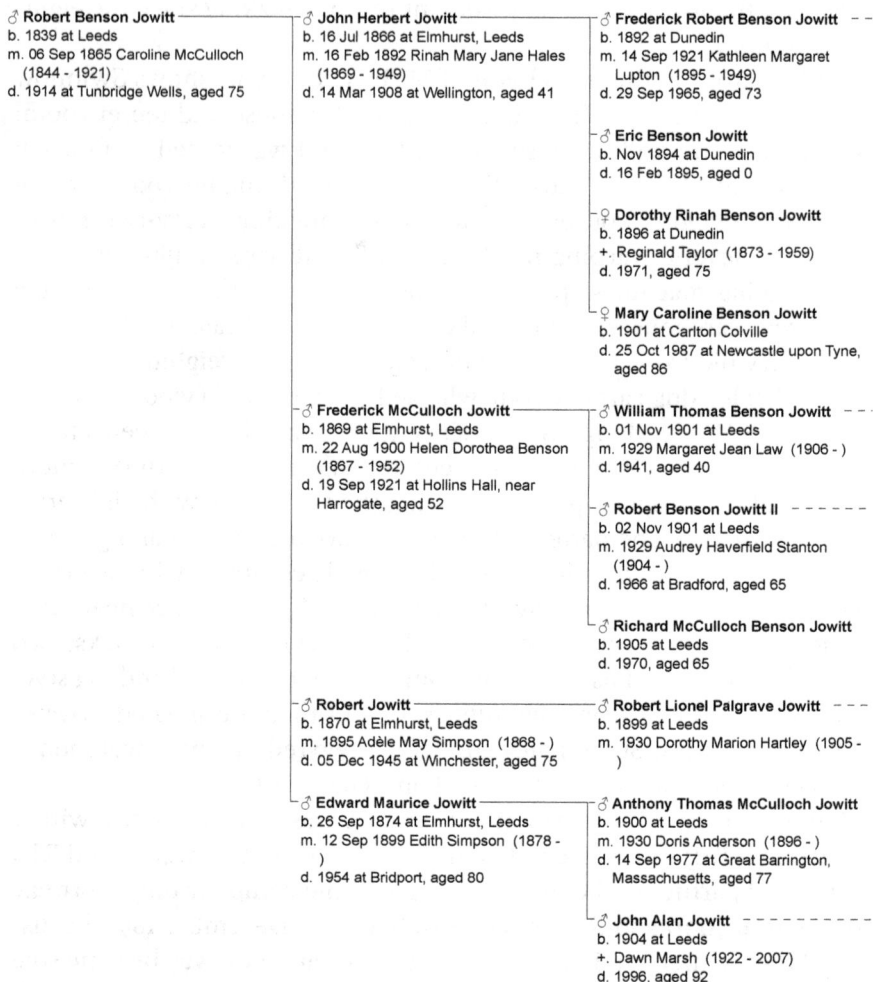

ROBT. JOWITT & SONS AT THE CENTURY'S END

The (incomplete) Robt. Jowitt & Son branch profit and loss accounts[280] in Appendix III give some idea of financial performance of the partnership at this period. The company as a whole made a loss of £74,102 in 1900, recovering to a profit of a mere £39 in 1901, £15, 336 in 1902 and £12,974 in 1903.[281] According to RBJ, the 1900 losses were light on 'lead' debts but heavy on stock losses.[282]

In 1900, RBJ made a rather gloomy analysis of the company's performance under his leadership over the previous twenty years. Excluding the years 1888, 1889 and 1899 which he described as exceptional (but for which, unfortunately, the figures are missing) and underestimating 1900's loss at £63,000, the profits and losses had exactly balanced after allowing 5% interest per annum on capital invested.[283] The profit and loss figures are given in Appendix IV.

What RBJ left unsaid, and what was felt by many members of the Jowitt family before and since, was that, whether the company was profitable or not, it was their purpose in life; it was their duty to provide employment in Leeds and elsewhere; it defined who they were and, in those more religious days, it was their sacred mission.

It was also in 1900 that Robert Benson Jowitt and his wife Caroline retired to Tunbridge Wells, Kent, where the air was cleaner and the climate milder than that of Leeds. He died in 1914 leaving Caroline an annuity of £1,500 to be paid in two parts, on 10 March and 10 November each year, from the income of his share of the partnership. His sons were named as joint executors and trustees.

The Sons of
Robert Benson Jowitt

Robert Benson Jowitt had four sons, John Herbert ('Jack' born 1866), Frederick McCulloch (b. 1869), Robert II (b. 1870) and Edward Maurice ('Tom' b. 1874). All four were sent to leading public schools – Jack to Charterhouse, Frederick[284] and Maurice to Marlborough (where Maurice excelled at cricket), and Robert to Radley. He offered each son the choice after school of either going to university or travelling around the world. Jack and Tom chose university while Fred and Maurice chose to travel the world.

JOHN HERBERT JOWITT

Jack continued his studies at Hertford College, Oxford, but sadly all that he learnt at Oxford was a taste for a wide range of sports[285] and gambling, and he left under a cloud without taking his degree. He was then despatched to New Zealand, the remotest and possibly the most difficult corner of the Jowitt trading empire. Jack returned to England briefly, and he and Fred were admitted partners in the firm in 1890, along with W. M. Tate and John Thomas Gibson, the company's first Melbourne Manager.

Jack returned to New Zealand in August 1890 and a new company, J. H. Jowitt & Co., was set up in October 1892 with credit of £25,000 from the Bank of Australasia.[286] It was later arranged that it should be paid commission of 1½% once wool was on board ship and a further 1½% at an unspecified date, presumably after the wool had been received and checked.[287] Although Robt. Jowitt & Sons had imported 3,442 bales from New Zealand in 1888,[288] Jack's new company seems to have left little evidence of its existence, other than a few small shipments such as that of 186 bales in 1894, and an advertisement for Kangaroo Island Guano, proclaiming it to be 'EQUAL TO ANY GUANO I HAVE USED ON TURNIPS,' which lists J. H. Jowitt & Co. as a distributor.[289]

On 16 February 1892 Jack married Rinah Mary Jane Hales, daughter of the late Samuel Hales, one-time co-proprietor of the Blue Spur gold claim.[290] The service was held at All Saints' church, Dunedin. Nestling among the trees, the church was 'a picturesque building, and seemed

Fig. 28 Rinah Hales and Jack Jowitt.

to lend itself to the romance of a wedding'. It was an opulent affair, the bride appearing fashionably late, and

> An audible murmur of admiration was heard as she passed along the aisle attended by her six bridesmaids. Her gown was exquisite, of rich white duchesse satin and silver brocade. The bodice was in the full empire style of duchesse satin, with high collar and puffed sleeves, satin and point d'esperit [sic] lace quaintly folded about the front. The bodice was also edged with a deep flounce of the same lovely lace. The petticoat, of duchesse satin, was quite plain, edged with a flounce of point d'esperit [sic], caught up at intervals with true-lovers' knots. The train was of silver brocade, in the Watteau style. falling from the shoulders, and glinted and glistened with every movement. A spray of orange blossom on the corsage and another on the train completed one of the handsomest wedding gowns that has been worn in Dunedin.[291]

Apparently, the groom was also present, for the same report tells us that

> After the usual congratulations the newly-wedded couple left amid a shower of rice and rose leaves, en route for the lakes, where they spend their honeymoon, on their return leaving again for England on a lengthy visit to Mr Jowitt's people.

Their visit to England does, indeed, seem to have been lengthy, as their return journey only began when they embarked on the RMS *Coptic* at London on 4 August 1892.

For the next few years, Jack and Rinah lived what appears to have been an idyllic life. Rinah's position at the very centre of the Dunedin smart set is documented in the local newspapers and, whatever his father expected him to do out there, Jack seems to have found plenty of time for leisure activities. A golfer[292] and cricketer,[293] he also enjoyed yachting and was on the committee of the Dunedin Jockey Club[294] and the Otago Rugby Football Union.[295] He was a stalwart of the local Dog Society and in one year at the Southland Show we learn that 'his team of retrievers won everything in the black retriever class,' that his Heather Pluck was runner up in the Best in Show and he won in the Rough Collie Bitch category with René, the Rough-Coated Collie Bitch Winners' class with Rimu Brighteyes and the Rough Collie Dog category with Rannoch.[296] The dog-breeding seems to have been more than a hobby, as Jack put his dogs out to stud at a charge of between three and fifteen guineas a time.[297] Somehow, Jack also found time for amateur dramatics, appearing in an amateur production of the Crimean War drama, *Ours,* at the Timaru Royal.[298] All this was very well, but could such a lifestyle be financed from the proceeds of the wool, fellmongery and guano trades? Apparently not, for RBJ noted in February 1897,

> Very disappointing and grievous letter from Jack this morning. I fear he is hopelessly insolvent. I write him very plainly, especially as to the necessity of absolute candour. It seems clearer than ever that he had better give up wool and fellmongery and take a small farm for breeding horses, etc.[299]

Six months later, it seems that no remedial action had been taken. When Fred and Maurice stopped off in New Zealand during their world tour in August, they quickly ascertained that Jack was indeed insolvent, and Fred wasted little time in reporting the situation to his father:

> A telegram from Fred from Dunedin, giving some words out of an 'animal' code. It confirms all our worst fears, adding that Rinah now knows all the circumstances. We wait in suspense for letters and further news about Jack's future. It must have been a painful time for Fred and Mott [Maurice].[300]

... Thought much about Jack's future in evening. Possibly an agency in the West [of England] might be the thing for him.[301]

On Fred's advice, Robt. Jowitt & Sons did not accept a New Zealand draft against a shipment which he believed Jack had drawn under pressure from the bank. Since Jack had exhausted his credit limit with the partnership, RJ&S was within its right to do so. As the full extent of the problems became apparent, RBJ took steps to ensure that Jack's financial recklessness did not taint the firm, giving instructions that no notepaper was to be used with 'Dunedin' on it. The situation had serious implications for Jack's immediate and wider family.[302] RBJ insisted that Rinah and the children should come 'home' to England (though they had lived all their lives in New Zealand) while Jack wrapped up his business affairs as best he could before following them:

Very bad news from Dunedin and the cablegrams have been most bothering. Rinah and children are on way home, having left by 'Gothic' on 19th, due Plymouth Sept. 28th. Yesterday at great cost to myself I wrote a full account of the whole affair to Mrs Hales at Dinan. We sent Fred by cable another cash credit £1500, in addition to the previous £1000, and also a wool credit £2000 as wired for.[303]

Rinah's mother attributed the catastrophe to 'horse-racing and Mason'.[304] This is possibly a reference to Mr R. J. Mason, a familiar figure in the local racing set.

After his return, Jack was briefly placed in charge of the West of England Branch. It was not a great success and he was not made a partner when the company was re-organised in 1900. He was asthmatic and his doctor recommended that he return to New Zealand, which he did in 1908. Sadly, he died onboard the ship from Australia to Wellington. He was buried in the Northern Cemetery, Dunedin, next to his son, Eric. The tombstone reads:

In
Loving
Memory
of
John Herbert Jowitt
Born at Leeds, England

16th July 1866
Died at Wellington
14th March 1908
Aged 41 Years
Also Eric
Infant son of the above
Died 26th Feb 1895 Aged 3½ Months

It is said that, towards the end of his life, RBJ felt that he had treated his son rather harshly. Jack was survived by Rinah, a son, Frederick Robert Benson Jowitt, and two daughters, Dorothy and Mary Caroline ('Moya'). Rinah and her children later returned to New Zealand and were sustained by an allowance of £500 per annum from RBJ, a sum which Rinah seems to have found inadequate. Her son, Frederick R. B. Jowitt, expressed a strong desire to become an engineer, but this was vetoed by his uncle, Fred McC. Jowitt, who, perhaps believing that Rinah was rather extravagant, seems to have wished to keep the family as far away from England as possible:

I am your executor and the children's guardian and a great responsibility rests upon me to advise you right...

First Fred... unless Fred turns out far above the average – in fact almost a genius – there is very little chance for him as an engineer in England... the profession is overcrowded...

I know that he has always talked about his becoming an engineer and he looks forward to the work but the practical thing is whether Fred can earn enough to keep himself and make a home for you and the girls...

I have decided to give him a chance in the wool trade... with a partner called Mr [Thomas] Beaumont. This is quite separate from the home concern although the title is the same... If he gets through a good training he will be able to earn a good living for himself... he will find the wool trade a far better paid... opening. Jack would have agreed...

... there is really no opening here with all these young people (my 3 boys, Tommy's 2, Bob's 1) coming on and in Australia he has an open field...

Now with regard to Dorothy and Moya. I feel strongly that they will have a more open healthy and freer life in Australia than they could possibly have if you lived in the suburbs of London. The society you would get in Melbourne or Sydney would be a freer and more pleasant one than is possible in an overcrowded city like London. They would meet people of their own standing... and would be more likely to make happy marriages in Australia than they would in our conventional and overcrowded society...[305]

Frederick went on to say that he was suggesting to Rinah's mother that it was her 'clear duty' to move to Melbourne with her and contribute to the expenses of running the home.

When RBJ died in 1914, Jack's family were largely excluded from the will, with the result that his son F. R. B. Jowitt and his descendants held few shares in the company. Nonetheless, F. R. B. Jowitt came over to England and distinguished himself in the First World War. He also achieved his ambition, studying Textile Engineering at Bradford Technical College and becoming an invaluable part of the Bradford company as Manager of Hollings Mill, and eventually Chairman.

FREDERICK McCULLOCH JOWITT

Family legend has it that Frederick studied at the University of Heidelberg, but the records show that he did not.[306] It is probable that he studied at another German university. His role was so central to the company that his story will be told in a separate chapter.

ROBERT JOWITT II

Robert attended Trinity College, Cambridge (admitted 1890, graduated BA 1893). While there, Robert Jowitt II and another Trinity student, Julius Simpson, were rusticated for the very Wodehousian offence of knocking off a policeman's helmet. Robert II feared that his father would not approve of his youthful exuberance, but Julius assured him that *his* father, Dr Reginald Palgrave Simpson, would take a more relaxed attitude, and so the two went down to the Simpson family home in Weymouth.

The Simpsons were a very musical family. Reginald's wife, Maria Georgina, was the daughter of Sir Julius Benedict, a highly-regarded pianist, conductor and composer who had accompanied Jenny Lind, the 'Swedish nightingale,' throughout her American tour in 1850. The Simpson and

Fig. 29 Robert Jowitt II and Adèle May, *née* Simpson.

Benedict families had first met through their involvement in the Norwich Music Festival,[307] at which Benedict conducted every year between 1845 and 1878. Sir Julius had studied under both Johann Nepomuk Hummel and Carl Maria von Weber and was regarded as the latter's most talented pupil. Reginald's brother, the dramatist John Palgrave Simpson, translated Baron Max Maria von Weber's biography of his father into English in 1865, possibly at Sir Julius's suggestion.

Fig. 30 Robert II's wife May with Frederick McC. Jowitt (in boater)
and his schoolfriend A. T. Keeling.

While staying with the Simpsons, Robert II met Julius's sister, a pretty, golden-haired young girl called Adèle May (known simply as May). He married her in Weymouth on 3 January 1895 and, on the following day, the couple set out for their honeymoon in Dresden where May had studied music. Their honeymoon nearly ended in disaster when, en route to Dresden, the paddle steamer *Empress* suffered a mechanical fault as she was about to enter Calais in a heavy gale.[308] She was slammed against the west pier, losing her bridge and starboard paddle box. With the *Empress* beginning to take on water and drifting helplessly westward, the passengers seem to have shown remarkable *sang froid*:

> Passengers say that the collision, which shook the vessel from stem to stern, produced some consternation on board. One gentleman put on a lifebelt by way of precaution; but his example was not followed by his fellow passengers.[309]

Fig. 31 Edward Maurice Jowitt and Edith *née* Simpson.

Fortunately, the storm forced the ship onto the relative safety of a sandbank, and all the passengers escaped unscathed.

Between 1900 and 1919, Robert II often deputised for his brother Frederick. Although a Director of the limited company from its incorporation, he played a relatively minor role in its management.

EDWARD MAURICE JOWITT

Maurice, variously called Edward, Mott or Tom, followed Robert's example, marrying May's equally pretty younger sister, Edith Georgina Palgrave Simpson, at Melcombe Regis, Weymouth, on 12 September 1899.

For many years he ran the scouring and carbonising section at Cliffe Mills. He was an active Director of the company for many years, serving as Chairman between 1921 and 1939. He managed to combine these duties with a comfortable country lifestyle, making his home at the elegant Strode Manor, Netherbury, West Dorset.

Frederick McCulloch Jowitt
(1869–1921)

'as starch as a Quaker'
John Sheffield, 1740[310]

Although John Jowitt I had apparently been involved in cloth manufacture in the eighteenth century, the Jowitts largely restricted themselves to the traditional wool stapling and consignment business during the nineteenth century. The exceptions were Robert Benson Jowitt's purchase of the G. R. Portway & Co. woollen and worsted mill, and the William Gibson fell-mongery, scouring and carbonising branch at Meanwood (which was demerged in 1906, with Jowitts walking away from UK fellmongering for some years).[311] There were clear dangers in such a restricted operation – if wool prices fell, as they frequently did, the staplers would suffer, but their customers, the manufacturers, would benefit. Moving into processing the wool offered obvious advantages but required substantial capital.

For Robt. Jowitt & Sons the capital needed to expand into manufacturing came through Frederick's marriage. On 22 August 1900, he married Helen Dorothea ('Nellie') Hope, daughter of his great-uncle, W. T. Benson and widow of Charles Cowan Grant Hope. Her late husband, of the Montreal wine merchants and grocers, John Hope & Company, had been appointed a Director of her family's company, the Edwardsburg Starch Company, in 1894 and his family firm granted sole agency for ES products to the grocery trade throughout Canada.[312] Hope's over-fondness for alcohol led to his death in 1898 at the age of thirty-seven, leaving his young widow Helen as rich as she was beautiful, having inherited considerable wealth from both her father and her first husband. While there was nothing stiff about Frederick, he owed much of his visionary expansion of the company to starch.

Frederick and Nellie had three sons, the twins William Thomas Benson Jowitt (the older by twenty minutes) and Robert Benson Jowitt II, born 1901, and Richard McCulloch Benson Jowitt, born 1905. The twins were far from identical; William was a good-looking child and grew up to be a handsome man whereas Robert was not a particularly attractive child whose poor eyesight required the use of thick spectacles. Unfortunately,

Fig. 32 Frederick McCulloch Jowitt (1869–1921),
first Chairman of Robt. Jowitt & Sons Ltd.

Fig. 33 Helen Dorothea Hope *née* Benson who married her cousin
Frederick McC. Jowitt in 1900.

neither their mother nor their nanny could conceal a partiality for the handsome twin, a fact which did not escape Robert's notice. This left him with something of an inferiority complex which cannot have been helped by being told that his brother, being the eldest by twenty minutes, was destined to be Chairman of the company. In later life, his feelings of inferiority may have contributed to his binge drinking. The third son, Richard, was generally considered a little unstable, and was given to gambling.

It was in 1900 that the company acquired the woolcombers, S. & S. Musgrave, taking them into the high-end market of supplying tops to the worsted trade. The two companies seem to have had a long acquaintance, both having been involved in the Cotton Districts Relief Fund in the 1860s, when S. & S. Musgrave was based in Pond Street, Leeds.[313] Sam Harland joined Jowitts from the Musgrave company, starting an association between the Harlands and the company which lasted more than seventy years. Herbert Lee also joined Jowitts from Musgraves, remaining until his death in 1925. During Frederick's tenure as senior partner the firm took offices in Swan Arcade, Bradford, although it did not vacate its Albion Street headquarters until after it bought Hollings Mill.

With the oldest brother, Jack, showing little interest in the business, Frederick became the senior partner when his father retired on 14 December 1900.[314] His younger brothers, Robert II and Tom, were admitted partners in the company on 17 December 1900, along with William Gibson. It was a very difficult year for the young Jowitts to take the helm, with the company suffering a massive loss of £74,102. Although both Tom and Robert II played an active part in managing the company, Frederick, as senior partner, was very much in control. His fifty percent share of the profits (or losses) goes nowhere near reflecting his share of the workload and responsibility. Indeed, it was only during his rare absences abroad[315] that his brothers would deputise for him.

Just as his father and grandfather before him, Frederick reinvigorated the family firm. Within a few years, as well as the merchanting branch and branches in London, Leeds, Halifax, the West of England (operating from Trowbridge, Wiltshire) and the various overseas operations, the company had a scouring branch at Carr Mills, Leeds, a scouring and carbonising operation at Cliffe Mills, Great Horton, Bradford (managed by A. R. Hummel,[316] whose father was Professor of Dyeing at the Yorkshire College, and then by E. M. Jowitt), a combing operation at Try Mills, Bradford (moved to Hollings Mill, Sunbridge Road, Bradford, in 1909), and the William Gibson fellmongering operation in Meanwood, Leeds. More or less by accident, the company had also acquired a Colonial Produce Branch, of which more will be said later.

In 1901, William Gibson & Co., the fellmongering operation, bounced back from a loss of over £10,000 the previous year to a profit of £5,410, and the West of England Branch made an acceptable £588. However, other parts of the company recovered very slowly as they re-oriented their buying to support the new combing operation. East London made a mere £115, Port Elizabeth £326 and Melbourne a loss of £23. All this was very uncomfortable for the branch managers who relied heavily on profit sharing agreements. The profit for the whole company in 1901 was a mere £39.

By 1902, the branches were making more substantial profits with West of England making £1,553, William Gibson & Co. £9,062, East London £8,492, Port Elizabeth £2,874, and Melbourne £2,704.

Profits declined in most branches over the next few years and fell steeply into loss in the East London Branch under Victor Tate (of whom more in the next chapter), the Sydney Branch under John Gibson and the Melbourne Branch under T. Beaumont. In Melbourne and East London, the losses were attributed to over-paying for wool in the local markets. In stark contrast, the Bradford topmaking/combing operations steadily increased in the years leading up to the First World War. In 1909 the Combing Branch at Try Mills made a profit of £1,958 3s 11d, the Scouring and Carbonising Branch at Cliffe Mills £1,506 6s 6d, and the Bradford Top Department £25,504 4s.[317]

The early part of the twentieth century saw a steady growth in the overseas customer base, especially the supply of wool to the growing Japanese market. Among the new customers were the China & Japan Trading Co. Ltd (later the China, Japan & South American Trading Co. Ltd), Mitsui & Co., the Tateichi Trading Co., the Achida Trading Co. Ltd and Okura & Co. Trading Ltd.[318] Of these, Okura was by far the largest of the company's oriental customers, if the number of samples supplied to them is any indication. Okura-gumi, as it was known in Japan, was a highly diversified *zaibatsu* (conglomerate) and the first independent Japanese company to set up an office in London, in 1874. It became the largest Japanese customer of Jowitts in Australia and South Africa and was a major customer for tops,[319] a situation which lasted until the Japanese installed their own combing operation in the 1930s.

The move into combing with the purchase of S. & S. Musgrave (owners of the Try Mills combing plant) transformed the company from wool merchants into a concern which made most of its profits as topmaker combers. A move from their longtime Head Office in Albion Street, Leeds, to Bradford, a city so synonymous with top production and the worsted trade that it was known as 'Worstedopolis', was only natural. Frederick McC. Jowitt moved his own office to the Bradford combing mill in the first few years of the twentieth century, and in 1911 the company bought Hollings Mill, Bradford, which it was already renting. Part of it was converted into a plain but comfortable Head Office. The Managing Director's office, with a large cast-iron pillar supporting the ceiling, looked more like the captain's cabin in a nineteenth-century warship than the hub of a successful multinational business.

Rabbits and Butter

To speak quite plainly about Mr. Edwards, who has been with us for 30 years, he is, I admit, somewhat unbusinesslike in details, but with clerical supervision we get over this and, on the other hand, I am quite convinced that he thoroughly understands the selling of rabbits, and knows well the somewhat tricky and curious customers he has to deal with.

Frederick McC. Jowitt, 1906[320]

By the early twentieth century, Robt. Jowitt & Sons had become a truly multinational company, with branches in Bradford, London, Melbourne, Sydney, Port Elizabeth, Durban, East London and Cape Town. Nonetheless, it was still very much a family business and, since Robert Benson's retirement in 1900, almost the complete burden of controlling this far-flung empire lay on the shoulders of Frederick McCulloch Jowitt ('Mr Fred' as he was often referred to in correspondence). In an age before international air travel and when there were no telephone links between Britain and Africa or Australia, it was inevitable that the branches would be given some autonomy, and it was also inevitable that this would sometimes have unfortunate consequences.

Despite clear instructions from the head office in Bradford, the Melbourne office bought wool at the top of the market, saddling the company with large losses. Frederick would always try to deal considerately with the company's agents and managers, while pointing out their errors. Since they received commission on top of their basic salaries, the managers had every incentive to learn from their mistakes. When the market was bad, the local managers were allowed to overdraw on their accounts, running up debts with the company which they had to be gently cajoled into repaying, often over a number of years.

VICTOR TATE

As early as 1896, it was clear that there was a problem with W. M. Tate's son, Victor, the young, enthusiastic and incorrigibly naïve manager of the East London branch. RBJ wrote to him on 23 December of that year,

I need hardly tell you how astonished we were to learn by Sheard's letter of 15 Nov. that you had been invoicing a considerable portion of your purchases above actual cost.[321]

Remarkably, RBJ accepted that Victor had 'no intention to defraud' and he remained in his post after RBJ's retirement in December 1900, thus becoming Frederick's problem.

While many managers owed the firm one or two hundred pounds, Victor (who was on £300 p.a.) owed it £4,679 by the beginning of 1905, including money they lent him to buy an orange farm. On 11 January, Frederick wrote to Victor:

... I wish particularly to draw your attention to the promise you gave me that your drawings in future should not exceed in one year more than £600... I sincerely hope that very soon you will be able to redeuce [sic] this overdraft, and to put your affairs straight.

You will, of course, also bear in mind your solemn promise not to undertake any more outside speculations.

Please let me have a line from time to time as to how your private affairs are progressing.

This season is commencing under very difficult circumstances, as wool is selling with you far above the level of our market, and it is impossible to give you a free hand either for combing wools or for wools for resale... [322]

Meanwhile, Victor had tried to renegotiate the particulars of which mortgages (on the orange farm and other properties) were offered as surety for the loan with John Sheard, the manager of the Port Elizabeth branch (and, it would seem, in overall charge of the South African operation). Since these arrangements had been agreed the previous September and had still not been implemented, an exasperated Sheard recommended that a more stringent bond should be demanded from Victor. The firm, however, was willing to retain the original agreement, and wrote on 10 February pressing Victor to fulfil the terms:

We must therefore ask you, without further delay, of any kind, to press forward all arrangements, and we have instructed Mr Sheard to see that it is done. We have no doubt that we shall soon hear that this is so. [323]

However, Victor's affairs did not improve and Frederick, in a letter of 10 August 1906 to John Sheard, described Victor's position as 'a very serious one'. He agreed with Sheard's assessment that Victor's income would not be sufficient to service his debts. Despite this, Frederick intended to delay a resolution of the matter until he was able to talk to Victor when he came over in 1907. In the meantime, the company would continue to finance him. Writing to Victor on the same day, Frederick stressed his wish to help him. By 21 September, he was sanctioning him to overdraw by a further £100–150.[324]

Unfortunately, Victor's performance in his job also gave cause for concern. He bought far more short wools than the Bradford branch wanted. Having ordered five hundred bales, Jowitts managed to use up a thousand, but the East London branch was still left with stock which was worth less than its purchase price.[325] Despite this, Frederick agreed with Victor that he would be allowed to withdraw up to £1,200 in 1908.[326] However, matters came to a head before the end of 1907, when Victor continued to buy in a falling market. On 27 December, Frederick wrote to him:

> From my former letters, you will realise now only too well what a terrible mistake you have made, and how you have lost money for us and more or less compromised your future.

> I am writing this letter much more in sorrow than in anger, for as you know I have always appreciated your efforts to improve and advance the interests of the firm, and I have a personal regard for you.

> I have just now carefully read over all your business letters and your private letters to me, and the cheerful optimism and reckless energy clearly expressed in all your public and private letters, as also in your cables, is in grim contrast to the actual realities of the situation.[327]

Not only had Victor overbought, he had bought indiscriminately. Frederick complained that much of the wool he had bought was not up to standard, adding 'We do not want this kind of buying, and we do not intend to have it in the future.' Victor had mortgages of £10,000 on land worth just that amount and Frederick could see no way that he would be able to service the debt on his salary and commission. His debt to Jowitts was still over £4,000. Despite everything, he offered to keep Victor on for another year at a salary of £500 and allow him to overdraw by a further £300.

However, further investigation by Frederick found that Victor had been buying wool at up to 2d per pound above the limit set by head office, and even more of it than he had previously suspected was not up to the specified standard. Frederick remarked wearily,

> For some inexplicable reason, you cannot buy wool quietly, carefully, or in any way near your limits. In fact your buying is more speculation and unless the market advances considerably, every bale you buy leaves a loss, and this result is all the more striking when compared with your fellow managers in South Africa.[328]

In the light of this, and anticipating a continuing fall in prices, Frederick closed the East London branch and finally dismissed Victor, allowing him to draw a further £100, £60 of which he would send directly to Victor's wife in London. After all this, Frederick still expressed his deep regret, and he was delighted to receive what he considered a 'plucky and straightforward' letter from Victor, saying, 'it confirms the good opinion I have always had of you'.[329] The orange groves, on which Jowitts held mortgages against Victor's borrowing, were later washed into the sea.

Perhaps at this point Frederick realised that he had allowed friendship to cloud his judgement and it was not he but Sam Harland who took the decisive step of sending Donald Bertram Sykes from Bradford to sort out the mess. A schoolmaster's son with a desire to travel (although he had his eyes on South America rather than South Africa), he came armed with a power of attorney despite being below the age of majority. Sadly, Victor shot himself while Sykes was staying as a guest on his farm.[330] The only good thing to come out of the affair was that in D. B. Sykes the company gained a very able manager who continued to work for them in South Africa for a further sixty years.

MR F. M. EDWARDS

Other branch managers also exceeded their authority, and Frederick always showed great forbearance when dealing with them. Sometimes, it must be said, his leniency seems to have been inexplicable. One such case was Mr F. M. Edwards, manager of the Colonial Produce sub-branch in London. Frederick had been persuaded to import Australian butter and rabbits on consignment by Edwards and by R. B. McComas of the well-known wool-brokers, William Haughton & Co. (the supplier of these items, with a rabbit-freezing plant at Mount Gambier, Southern Australia). Frederick was wary of entering a business he didn't understand, but he was convinced that Edwards understood the rabbit trade and the 'somewhat tricky and curious customers' – no doubt he was thinking of the human kind, as the rabbits were already frozen. He also believed that it would be a way of providing Edwards with a healthy commission and of subsidising the London office. By imposing stringent procedures on the new business, Frederick overcame his own reservations and those of his partners.[331]

The trade in both rabbits and butter turned out to be disastrous, and Frederick discovered far too late that McComas had been right to be concerned at Edwards's unbusinesslike attitude, which led, in the first place, to him selling one consignment of rabbits twice, a mistake which cost the company £400.

Frederick had been alert not only to the rather tricky customers involved in the rabbit trade, but also to the dangers of dealing in any area, such as the butter trade, in which he did not have personal contacts. Therefore, he gave Edwards explicit instructions to protect the firm against potential default or fraud. He recognised that they needed someone with expertise in the butter trade and Edwards introduced a wholesaler called Welch with warehousing in Tooley Street. He stipulated that butter was only to be purchased from McComas, whom he knew and trusted, and that it was to be supplied to Welch only up to a value of £100 at a time. Edwards was not to supply Welch with more if he had built up a debt of more than £300–400; that way, the firm was taking little risk. Events were to prove the importance of these precautions and also Edwards's inability to follow orders.

Frederick discovered too late that Edwards had provided Welch with £300-worth of butter in March, £3,000-worth in June and £700-worth in August and September without receiving any payment from him. Welch had promptly sold the butter and was by then bankrupt (indeed, according to Frederick, he had been so for a year). This was particularly shocking as the stock reports from Edwards had indicated that Robt. Jowitt & Sons still had some £4,000-worth of butter stock, although all this had been sold to Welch. The company lost £3,000 on this deal alone and, as Frederick pointed out to Edwards, fifty percent of this came straight out of his own pocket as senior partner in the firm. Frederick concluded that legal action against Welch (whom he considered a crook) would have yielded no money and would have made them a laughing-stock of the commercial world.

Edwards also failed to follow William Haughton & Co.'s instructions regarding the sale of rabbits supplied on consignment, leading to a claim for £4,000 and a long and unhappy dispute between the two companies.

Understandably, Frederick wrote Edwards a furious letter, concluding by saying, 'we have always had and still retain, absolute confidence in your integrity and honesty – in fact this is your only asset,'[332] but even then he did not sack him, hoping to continue the Produce sub-branch, after a transition period, under the managership of Mr Charles R. Valentine, who had been Produce Commissioner in London for the governments of New Zealand and New South Wales,[333] and had worked for Nelson Brothers and the Colonial Consignment and Distributing Company before running his own import company, Valentine & Co. Ltd, which was wound up in 1902.[334] However, after two very difficult years in the wool trade, Frederick decided that the strain of carrying out such a large and a diverse business was too much for him. In 1908, he announced that he was closing the sub-branch to concentrate on the Bradford top-making operation.

It might not seem that a small sub-branch running out of the London office could cause much of a distraction to this large multinational company, but this would be to misunderstand the size and nature of the operation. The Colonial Produce Department was certainly not a major money-spinner but it had potential and, in a way, it seemed to make sense; they were using their network of overseas and local contacts and their logistical skills to build a totally new operation which could augment the wool business. Yet, when one examines the range of produce handled and the number and geographic spread of its customers, it is clear that it must have consumed enormous staff and management resources. The branch accounts show that produce included rabbits, hares, chicken, bacon, eggs, cheese, butter, salmon, quails, oranges and apples. Customers ranged from Sir Charles Petrie, former Lord Mayor of Liverpool, to J. Sainsbury and the Army & Navy Stores in London, A. Laurent & Cie. and R. Hochart in Boulogne, the River Plate Fresh Meat Company in Glasgow, F. Freeman in Exmouth, P. Molineux in Swansea, and a hundred or so other outlets in Leeds, Birmingham, Manchester, Nottingham, Ireland and France. Leeds even had its own Produce Department to help sell the imported goods.[335]

Having decided to close the sub-branch (but retaining a skeleton London office under Edwards's son Cecil), Frederick dispensed with Edwards's and Valentine's services. His letter to Valentine was polite but curt. He finished by saying, 'Our business relations have always been pleasant, and I regret they have to stop. It, however, must be so.'[336] In making Edwards redundant, he still offered him a pension of £60 per annum and, with remarkable generosity, to write off the £500 which Edwards owed the firm, adding rather sourly,

> I mention that you owe us this amount of money exactly the same as if I had handed you gold to that amount, and I cannot admit for a moment the fancy deductions you made from it in one of our late interviews, and I may add that before we definitely settle on the £60 pension I should like to hear from you that this is an exact and true statement of the matter.[337]

Edwards continued to receive a pension (increased at some point to £100 p.a.) up to his death on 20 March 1932.[338] Given this painful lesson, it might be assumed that Robt. Jowitt & Sons would never touch frozen rabbits, or any other meat, again, but in 1912 they were still making such shipments, and this enterprise was still causing them trouble and losses. They had sold the Produce sub-branch to Mr Valentine who turned it into the British Standard Produce Company Ltd, and it was through this company that Robt. Jowitt & Sons shipped the frozen meat. It appears that they kept a material interest in the business.[339] Unfortunately, when Jowitts attempted to collect a shipment from Union Cold Storage in 1912, they were told that the British Standard Produce Company (which was wound up in June 1912)[340] had failed to pay some outstanding debts. Union Cold Storage claimed that under their standard terms Jowitts were obliged to pay ten shillings per hundredweight on all the goods in the shipment, not only those belonging to themselves. They therefore paid Union Cold Storage the sum of £833 1s 4d in order to release their own goods, but instituted proceedings to reclaim what they considered to be the excess. The case came to court in 1913 and Mr Justice Scrutton considered that Union Cold Storage had a very strong case. However, he did not enter judgement, inviting the parties to come to an agreement.[341] Nothing further was reported, so it must be assumed that the case was settled out of court. It was the end of a brave but unsuccessful attempt at diversification. It would not be the last.

World War I

Very few of us realise with conviction the intensely unusual, unstable, complicated, unreliable, temporary nature of the economic organisation by which Western Europe has lived for the last half-century. We assume some of the most peculiar and temporary of our late advantages as natural, permanent, and to be depended on, and we lay our plans accordingly... Moved by insane delusion and reckless self-regard, the German people overturned the foundations on which we all lived and built.

John Maynard Keynes, *The Economic Consequences of the Peace*, 1919

Wars can be bad for business, as the English wool trade discovered during the American War of Independence, the Napoleonic Wars and the Franco-Prussian War. But they can also be very good for business, as the American royalist, Benjamin Thompson (later Count Rumford), discovered when he supplied uniforms to British troops during the War of Independence. Similarly, Jowitt's customer Okura-gumi benefitted from munitions sales to the Japanese government during the first Sino-Japanese War. In Leeds, the demand for army blankets during the First World War transformed Francis Tennant Varey from pot-hawker and rag merchant to prosperous factory owner. He was worth more than thirteen thousand pounds when he died in 1929.[342]

For Robt. Jowitt & Sons, the war brought several changes. Since the late nineteenth century, the company had ceased to rely on Bradford Old Bank for credit ratings and had turned instead to the Schimmelpfeng Institute in London. However, in 1916 British and Australian governments closed it down, concerned that a German-owned company had access to sensitive British commercial information. More seriously, access to imported wool was restricted by German naval activities. A rationing system for wool was brought in and Frederick was appointed Deputy Wool Controller under Lieutenant Colonel Willey to help allocate available stocks fairly. It was said that the appointment was made because Jowitts were believed by officials to be the only honest wool merchants in Yorkshire, a judgement which must have delighted the Jowitts and offended their rivals in equal measure. His office for the duration of the war was the Great Northern Victoria Hotel, Bradford. Frederick was appointed Commander of the Order of the British Empire in 1918 for his valuable services.

Robt. Jowitt & Sons' wage bill in the Top Department declined significantly, from £6,116 10s 3d in 1915 to £4,891 18s 11d in 1916, presumably indicating shortages of both wool and manpower. The wages for warehousemen and foremen rose slightly, but the wages for sorters fell. This decline was reversed in the following year when wages rose to £6,985 11s 11d. Nonetheless the wage bill during the war years was eclipsed by the heavy cost of insurance, which reached £21,041 5s 1d in 1916 and a staggering £35,560 9s in 1917 (presumably largely through the Government's War Risk Insurance Scheme). The wages rose again in 1918, to £9,262 2s 2d, the increase in these two years being entirely accounted for by the increase in sorters' wages. The wage bill plummeted to £2,011 10s 9d in 1919.[343]

At the beginning of the war it looked as though the effects on business would be catastrophic. In August 1914, Frederick wrote to William Gibson,

> All our orders have been cancelled, and we have been compelled to put Hollings Mill and Cliffe Mills on short time, and in a few weeks we shall have no work in either place. The same, I think, will apply to the fellmongery, and we wish to place all three places on the same footing.

> We have determined to stand by our people, and to spin out all the work we can, so as to give them half-employment, and when we cannot employ them, it will be impossible, of course, to simply dismiss them, and we shall have to assist them to the best of our ability...[344]

Such was the squeeze on profits that the company reduced wages in Leeds and Bradford and asked R. A. Gleeson in Port Elizabeth and G. A. Reid in East London to accept reductions in their salaries to £250 per annum.[345] In the end though, Robt. Jowitt & Sons, like Francis Varey, actually profitted from the war as its wool was sold to make uniforms. Frederick did not inherit his grandfather's pacifist philosophy and he had even been a reservist in the Leeds Rifles, The Prince of Wales's Own (West Yorkshire Regiment). However, he was acutely aware of the suffering caused by the 'War to End All Wars'; indeed, his own nephew, Lieutenant (later Captain) Frederick Robert Benson Jowitt of the West Yorkshire regiment, was reported killed in action by *The Times* on 15 July 1916. His mother, Rinah, received official notice of his death. Happily, like Mark Twain, reports of his death proved to be exaggerated and he

Fig. 34 Jack's son F. R. B. Jowitt, who was reported killed in action, with his mother and sisters, Dorothy (left) and Mary Caroline ('Moya').

returned home safely, though badly injured. It is possible that his identification was delayed by the fact that his dog-tag incorrectly showed his name as 'Jowett'. Later he would tell his family how he recovered consciousness on the Somme battlefield to the sound of groaning, only to discover that it was his own voice which groaned. Having trained as an engineer, he brought valuable practical expertise to the company, especially later, in his role as Manager of Hollings Mill.

The company's accounts show many generous donations to service and other charities during and after the war,[346] but this was not enough for Frederick McC. Jowitt who was deeply unhappy with the idea that the company might have profited from the carnage. Believing that the Russian people had suffered particularly badly, he had the Accounts Department calculate what might be considered the 'surplus' profit for the war period and wrote out a cheque for that amount in favour of their government (by then Bolshevik). Fortunately for the survival of the company, the cheque required the signature of a second partner, which was not forthcoming.[347]

Robt. Jowitt & Sons Ltd

Anothers humour will nothing allow
To bee more profitable then a Cow,
Licking his lips, in thinking that his theame
Is milke, cheese, butter, whay, whig, curds, and creame,
Leather and Veale, and that which is most chiefe
Tripes, chitterlings, or fresh or powder'd beefe.

Taylor, *The Praise of Hemp-Seed,* 1620

It is clear that Frederick had long seen the advantages of turning the family business into a limited company. On 7 December 1910, his school-friend, the solicitor Arthur Trowbridge Keeling,[348] sent him a memorandum on the practicalities of doing so.[349] That it had not been put into effect by 1914 may reflect the difficulty of coming to an agreement with all the partners. After that, no doubt the plan was put on hold for the duration of the War.

The long history of the Jowitt wool partnerships came to an end on 9 October 1919 with the formation of Robt. Jowitt & Sons Ltd under the chairmanship of Frederick McC. Jowitt. The other directors were: Frederick's brothers Edward Maurice (Tom) and Robert; Sam Harland, whose family had been brought into Robt. Jowitt & Sons when Jowitts-bought S. & S. Musgrave in 1900; Herbert Lee, also formerly of S. & S. Musgrave; and George Blackwell, the company secretary. Frederick McC. Jowitt's wife, Helen Dorothea, although not a director, was a major shareholder (see Appendix V for the initial allocation of shares), having bankrolled the partnership's expansion with Hope and Benson funds in the early part of the century.

Leather & Veale (affectionately known to the Jowitts as 'Tough & Tender') of East Parade Chambers, Leeds, were the company's auditors, with their partner, Gerald C. Veale becoming a director. The solicitors Trower, Still, Parkin and Keeling of 5 New Square, Lincoln's Inn, became the company's solicitors, on account of the lifelong friendship between Frederick and their partner Arthur T. Keeling (a fellow Old Marlburian). The Jowitts had moved away from Becketts Bank in favour of Bradford Old Bank during the late nineteenth century. This was taken over by United Counties Bank in 1907, and that, in turn, merged with Barclays in 1916. For this reason, Barclays were appointed bankers to the new limited company and provided substantial financing. As in the partnership days, this did not come without strings, much of it being secured against wool stocks and, from 1927, personal guarantees from the directors.[350]

The company had barely been incorporated when the scouring and carbonising plant at Cliffe Mills burnt down. At the second board meeting, on 21 November 1919, George Blackwell stated that the estimated cost was £50,000.[351] Fortunately, their insurers settled the claims (other than that for consequential loss) promptly and the board were so impressed that they resolved on 23 January 1920 to write and thank the three companies involved, Guardian, Guardian Eastern and British Dominion. Despite unresolved negotiations with Ramsdens about breaking the Cliffe Mills lease, the company bought the Highbury Works, Meanwood, for £32,500, including some surrounding land, and moved the scouring and carbonising operations there.[352] The buldings had been erected by Samuel Smith in 1857 but the famous brewery company had returned to their old brewery in Tadcaster in 1914 and rented out the premises to William Gibson & Co.,[353] a fellmongery company with which the Jowitts had been in partnership until 1905.[354]

Confusingly, the company also owned part of another concern in the area, Meanwood Fellmongers Ltd, which, in 1919 was reported to be processing 1,050 long-wooled skins per week.[355] In January 1922 it recommended to the directors of Meanwood Fellmongers that the company be wound up.[356] An Extraordinary General Meeting of Meanwood Fellmongers was then held at the Jowitt offices in Sunbridge Road on 31 May 1922. Two weeks later, a second EGM was held at which it was agreed to wind up the company voluntarily and appointing Harry Douglas Veale (of Jowitts' auditors, Leather & Veale) as liquidator.[357] On the same day, Robt. Jowitt & Sons bought the assets, goodwill and name of the company from the liquidator.[358] As a formality only (all creditors were paid in full), a creditors' meeting was convened on 4 July.[359]

From the start, despite a policy of selling forward, the company struggled to sell the wool they had bought. The Top Department was instructed to over-sell up to 1,000 packs (a pack is 240 lb, so this equates to about 107 long tons, 109 tonnes), but for their products, at least, there seemed to be sufficient demand, and they were authorised to buy up to 250 bales per week[360] from the Cape market. At the 2 June 1920 meeting the Board resolved to cease buying for all departments and to press customers to

Fig. 35 Sam Harland who worked for Robt. Jowitt & Sons from 1900 and served as Managing Director from 1919 to 1939.

take up their contracts. Robt. Jowitt & Sons Ltd would accept bills in payment, except from companies with heavy debentures.[361]

Despite difficult trading conditions, the company made generous charitable donations throughout 1920 and, in July, at the suggestion of Sam Harland, the board voted to give all employees paid leave for the week of the Bradford holiday fair known as Bowling Tide. It was the act of a company which not only cared for the welfare of its staff but also had complete certainty in its commercial future. No Jowitt was now heard to mutter, 'beware of the feast!'

In further signs of self-assurance, the company set substantial new salaries for directors (see Appendix VI) and senior management in October 1920 and summarily rebuffed Barclays' demand for personal guarantees on their lending, although they were later forthcoming.

Frederick McC. Jowitt had successfully overseen the transition of the family partnership into a limited company and, by the beginning of 1921, at the age of 51, he seemed set to preside over this increasingly successful enterprise for many years. A tennis game that summer was to frustrate that expectation.

According to family legend, Frederick was left with a blister on his hand after playing tennis. It was a trivial injury, but it became infected and Helen persuaded her husband to consult Laura Sobey Veale, daughter of

Dr Richard Veale and sister of Gerald Veale, of Leather & Veale, the company's auditors.

Leeds Medical School, on the board of which both Frederick and his father RBJ had served for many years, had rejected Laura Veale's application to study medicine in 1897 (and, in fact, did not admit any women students until 1910),[362] so, with the encouragement of Elizabeth Garrett Anderson, she had studied at London University instead, graduating MB in 1904 at the age of thirty-six.[363] She was the first woman doctor to be appointed to a Leeds hospital,[364] the Leeds Hospital for Women and Children. Her papers on 'the value of Harrogate spa treatment in relation to diseases of women'[365] and on colonic irrigation[366] suggest that she was at the less scientific end of the medical spectrum. Although this hardly makes her a quack by the standards of the time, she might not have been the obvious person to attend Frederick, especially since most of her medical experience was restricted to women and children.

Dr Veale treated Frederick using her favoured naturopathic remedies, but he died at his home, Hollins Hall, near Harrogate, on 19 September 1921 at the age of 52.

The causes of death, as given by Laura's father, Dr Richard Veale, on the death certificate, were Henoch's purpura and paralytic ilea. There must be some question about the diagnosis, as Henoch-Schönlein purpura, as it is usually known, is seldom seen in those over twenty. Even today there is no generally accepted treatment, but it is usually a benign, self-limiting disease, the greatest danger being the risk of kidney damage in about five percent of cases. In a very small number of patients, however, it can cause a blockage of the intestine, which would fit with the 'paralytic ilea' diagnosis, and this can be fatal. Fairly or not, there was always a feeling in the Jowitt family that Dr Laura Veale's treatment of Frederick had been inadequate.

At the board meeting on 23 September, Sam Harland described the unexpected death as a great loss. Frederick's old friend Arthur Keeling was appointed a director and Frederick's brother Edward Maurice Jowitt Chairman.[367]

Frederick's life had been insured by the company with two policies totalling £25,000, and this was paid out by way of further interim dividend free of income tax and, in accordance with article 3, the dividend was paid pro rata to the shareholders for the time being of the first 200,000 ordinary shares in the company.[368]

Of this sum, £1,475 was paid to Robert Jowitt II and E. M. Jowitt, being the proportion of this sum due in relation to the 11,800 shares[369] which they held as the two surviving trustees of RBJ's estate.

At the Ordinary General Meeting on 10 February 1922, it was resolved,

> that a final dividend of 5 per cent free of income tax be declared in respect of the year ending Nov 20, 1921, making, with the interim dividend of 5 per cent free of income tax already paid, 10 per cent free of tax for the year, and that such final dividend be forthwith paid.

Because RBJ's widow, Caroline, had died on 15 March 1921, this threw up the thorny issue of whether some of the £1,475 belonged to her estate under the Apportionment Act of 1870. This led to a court case to establish the correct application of the law. The judge in Jowitt v. Keeling, 1922, stated,

> Having regard not only to the absence of any mention in the resolution of November 4 of any period in respect of which the payment of the 25,000 pounds was made, but also to the nature and origin of the sum then distributed, and to the special rights in respect thereof conferred by art 3 upon the original holders of the first 200,000 ordinary shares, I am of [the] opinion that the payment of the sum of 25,000 pounds cannot properly be said to have been made for or in respect of any particular period so as to bring it within the scope of the Apportionment Act, 1870. Accordingly, I hold that the sum of 1,475 pounds is outside the purview of the Act, and is divisible among the residuary legatees to the exclusion of the estate of the annuitant. The costs of all parties (to be taxed as between solicitor and client) will he retained and paid by the plaintiffs out of the testator's estate.

While Frederick's death was undoubtedly a great blow, thanks to his work in restructuring the concern as a limited company with a sound management structure and secure finances, it did not cause the chaos and uncertainty which might otherwise have been expected. E. M. Jowitt's immediate appointment as Chairman (a position he held until 1938) ensured continuity and, under the terms of Frederick's will, his twin sons William Thomas Benson Jowitt and Robert Benson Jowitt II were nominated as directors at the board meeting on 10 January 1922, although, at the age of twenty, they were still too young.[370] They actually took up their positions on the board in 1926.[371]

The 1920s and 1930s

The policy of improving the foreign-exchange value of sterling up to its pre-war value in gold from being about 10 per cent below it, means that, whenever we sell anything abroad, either the foreign buyer has to pay 10 per cent *more in his money* or we have to accept 10 per cent *less in our money*. That is to say, we have to reduce our sterling prices, for coal or iron or shipping freights or whatever it may be, by 10 per cent in order to be on a competitive level, unless prices rise elsewhere. Thus the policy of improving the exchange by 10 per cent involves a reduction of 10 per cent in the sterling receipts of our export industries.

John Maynard Keynes, *The Economic Consequences of Mr Churchill*, 1925[372]

Even before the world had recovered from the devastation of the First World War, the seeds of future economic problems were being sown. Some European countries had borrowed heavily from the United States during the war, while Germany was saddled with a bill for £13 billion (some £550 billion at 2011 values) determined by the Inter-Allied Reparations Commission in 1921. Although the US government urged the European allies to curb their demands for reparations, they did not offer to modify their own demands for loan repayments. Britain itself had not accumulated too much debt during the war but her major export markets on the Continent were severely constrained throughout the twenties, and she endured a severe recession in 1920–21, followed by years of slow growth and high unemployment.

THE AUSTRALIAN OPERATION

Australian wool had been of major importance to the company in partnership days and the limited company put considerable resources into building it up. Key to this was Arthur Ainsworth Gibson, the Australian Manager (Director from 1926). Gibson, who was one of the chief appraisers for the Imperial Wool Purchase Scheme during the First World War, could almost have been said to have been born into Robt. Jowitt & Sons, as his father, John Thomas Gibson, had been Manager of the Melbourne Branch, becoming a partner before his sudden death at the age of forty-two in 1896. William Gibson, John Thomas's brother was proposed as a replacement,[373] but it seems unlikely that he actually went to Australia since he was fully occupied between about 1899 and 1906 running William Gibson & Co fellmongery in partnership with Jowitts.

A. A. Gibson was a familiar figure in the wool trade, both in Perth, where he was based, and throughout Australia. One Australian newspaper said of him:

A true English gentleman, of debonair manner, he possesses a beautiful tenor voice, which makes him in great demand at social functions. He enjoys great popularity amongst all in the wool trade.[374]

A. A. Gibson's older brother John had started working for Robt. Jowitt & Sons at the age of fifteen in 1896 and was made head of the Sydney office in 1906.[375] Due to unspecified trouble with John in 1925 (quite possibly indebtedness to the company)[376], A. A. Gibson was cabled with instructions to dismiss his brother and pay him a quarter's salary and passage home or £250 if he wished to stay in Australia. His powers of attorney with various banks were cancelled but, presumably at A. A. Gibson's urging, the company agreed to let John remain for a further trial period of three months.[377] He clearly redeemed himself as he was still working for Jowitts after the Second World War.

In 1927, at the request of A. A. Gibson, the company sent a promising young wool-buyer, Lawrence William ('Laurie') Firth, to help in the Perth Office at a starting salary of £350 per annum.[378] Sadly, he died less than three years later in a sailing accident when the yacht *Wattle* sank for no obvious reason between Rottnest island and Fremantle. Of the four men aboard, only one, a well-known solicitor, Gordon Bede D'Arcy, survived, having swum five miles to get help.[379] Firth's death was described by George Blackwell as 'lamentable'.[380]

THE SOUTH AFRICAN OPERATION

It is something of a mystery how the Jowitts, whose forbears Robert and John jr had been such staunch campaigners against not only slavery but also any exploitation of native populations, reconciled their consciences to the idea of establishing offices in that most exploitative country, South Africa. Nonetheless, this operation, under the able management of the Durban-based Donald Bertram Sykes (who was appointed a Director in 1924) and with W. R. Myers in Port Elizabeth, grew in importance during the years between the First and Second World Wars, especially in the supply of wool to the Japanese market. Even so, profits were tight and the Head Office made several efforts to reduce overheads. In 1938, D. B. Sykes offered to retire but the company chose instead to close the East London Office. The Manager, George Alexander Reid,[381] was offered and accepted the agency which he could run from home, with possible help from Durban or Port Elizabeth during busy periods.[382] The company agreed to pay him a commission on the following basis:

2% on first cost on wool and skins bought for the company's own account to a total value of £25000;
1½% on the next £12500;
1% commission on all purchases over £37500;
Commission on clients' orders of 1½%, and for Japan ¾%;
Commission on shipping only to be at the rate of ¾%;
and be guaranteed a minimum commission of £300 per annum for a period of three years.[383]

WALTER SCOTT & SONS – A 'WOOLLEN TRADE FUSION'

While the purchase of G. R. Portway & Co. in the nineteenth century had shown the very real risks of moving into new areas of wool processing, the successful absorption of the S. & S. Musgrave combing business in 1900 seemed to show that vertical integration had much to offer. The obvious next step was a move into spinning, and it was a proposed joint venture with Leonard Scott to buy a Kirkheaton spinning concern which brought Robt. Jowitt & Sons into a close and costly alliance with Walter Scott & Sons of Troqueer Mills, Dumfries.[384] Nothing came of the spinning proposal[385] and Jowitts soon dismissed the idea of such a venture. They later took a small step towards further integration by placing commission spinning of tops with Henry Illingworth & Sons at the rate of 10,000 lb per week.[386]

Walter Scott & Sons, an old established tweed manufacturer which had moved into other areas such as fancy worsteds, seems to have had ambitious expansion plans but limited liquidity. In fact, Scotts already owed Jowitts thousands of pounds. Their solution was another joint venture proposal. The usually cautious directors of Robt. Jowitt & Sons gave every impression of carrying out due diligence, only to rush blindly into an agreement which would have been fraught with difficulties even if it had not been launched in the face of a serious recession. The decision was even more remarkable since the company had been obliged to accept ten shillings in the pound on previous Walter Scott & Sons debt in 1901.[387]

After visits to Walter Scott & Sons and favourable reports by both George Blackwell[388] and Sam Harland (who proposed giving Scotts £30,000 credit),[389] the Jowitt directors were very amenable to the proposed joint venture.[390] The proposal, as laid out by Jowitts, was that Scotts would incorporate as Walter Scott & Sons Ltd with an ordinary share capital 75% owned by Scotts and 25% by Jowitts. In addition, there would be

£10,000 Debentures or Preference shares to Jowitts'

£30,000 Deferred Ordinary Shares to Ditto in respect of the amount the mill as is "written up."

£1000 each, but not free of income tax to the three Messrs Scott & 20% commission on the nett profits to the three Messrs Scott collectively.[391]

A subsequent report on Scotts described the company as 'as satisfactory as could be expected at the present time,'[392] hardly a ringing endorsement in the middle of a severe recession. Shortly after this, Scotts' indebtedness to Jowitts was estimated to be about £80,000.[393] Nonetheless, Jowitts did not hesitate to send a delegation consisting of Sam Harland, Gerald Veale, George Blackwell and A. T. Keeling, with Leonard Scott and a Mr Whitelaw, to discuss the joint plan with the Commercial Bank of Scotland.[394]

Other than the very high level of debt revealed by the proposed arrangement, the most worrying aspect was the joint Jowitts/Scotts bank account which, however carefully the terms were drafted, seemed to leave Jowitts very exposed (see Appendix XI). Surprisingly, the Commercial Bank of Scotland indicated their readiness to accommodate the arrangements (although on a temporary basis to bring down Scotts' debts)[395] and the joint venture then negotiated to buy the mills and machinery of the venerable Scottish concern, Robert Archibald & Sons,[396] manufacturers of 'tweeds and shirtings'.

At the same time that Robt. Jowitt & Sons was pouring money into the joint venture, it was continuing to examine Scotts' credit-worthiness on a regular basis. Although the warning signs were clear, the Board seemed completely oblivious to them. Sam Harland, after a visit to Dumfries in

January 1922 reported that 'he considered the old stock of Walter Scott & Sons was valued on a basis that it would be impossible to replace at even today.'[397] No-one asked why such ancient stock had not been sold, or whether it *could* be sold.

Even before the Memorandum and Articles of Association of Walter Scott & Sons Ltd had been received,[398] Scotts was asking Jowitts for help in financing the arrears in its manufacturing commitments, and £10,000 was duly advanced to the joint account.[399] Two months later, in July 1922, the contributions to the Joint Venture Account were already in excess of the amounts set out in the joint venture agreement, and a further £10,000 from Robt. Jowitt & Sons was being contemplated.[400] Nonetheless, the new alliance was trumpeted in the Scottish press (while remaining strangely reticent about the name of Robt. Jowitt & Sons):

WOOLLEN TRADE FUSION

What is believed to be the first occasion in the history of the woollen trade of a fusion of interests between Scotland and Yorkshire has taken place by the formation of a private limited company to take over the old-established business of Walter Scott & Sons, woollen manufacturers, Troqueer Mills, Dumfries. The Company has been formed in conjunction with one of the oldest of woollen merchants and top makers in Yorkshire. There will be no change in the management of the business.[401]

There is no indication that anyone at Robt. Jowitt & Sons was aware of the substantial debts owing to Scotts from Fraser & Wilson Ltd, a Dumfries woollen manufacturer and clothier, until 2 November 1922 when the minutes record,

The report of Messrs Leather & Veale was read over and it was Resolved to ask the Board of Walter Scott & Sons Ltd to adopt it and to insist on Fraser & Wilson Ltd carrying out the suggestions contained therein.

By February 1923 the Board of Robt. Jowitt & Sons were seriously concerned about the joint venture and decided that their Company Secretary, George Blackwell, should be appointed Chairman of Walter Scott & Sons Ltd. They also resolved,

That the loss arising from Fraser & Wilson Ltd be debited to the Scott Dormant a/c, the Joint a/c and the trading a/c of Walter Scott & Sons Ltd in such a manner that the Scotts pay to the uttermost of their ability.[402]

The Board also discussed rumours of irregularities in Fraser & Wilson's accounting with regard to book debts.[403] By April, the situation was bad enough for the company to give notice to terminate the joint account. Although it was still willing to trade with Walter Scott & Sons Ltd, it would only do so up to a credit limit of £30,000 running on three-month bills:

> The foregoing being subject to our having the right of inspection of the Company's books as at the present and also conditional on Scotts not obtaining credit elsewhere other than for mill supplies.
>
> Further all the foregoing to be conditional on the Company's banking facilities continuing as at present.[404]

At a special directors' meeting on 13 April 1923, attended only by Sam Harland, Herbert Lee and George Blackwell, it was decided to terminate the joint agreement entirely.[405] A joint meeting of the Boards of Robt. Jowitt & Sons Ltd and Walter Scott & Sons Ltd was then held at Jowitts' Sunbridge Road offices on 4 May 1923. It was clearly an awkward occasion, with Mr Whitelaw of Scotts asking whether the decision to close the joint account was final. Maurice Jowitt confirmed that it was.

Scotts agreed to try and work within the £30,000 credit facility. Leonard Scott pointed out that this would mean supplying no more cloth to Fraser & Wilson 'other than on the basis of prompt cash'. Given that Fraser & Wilson (which they were in the process of liquidating) already owed Scotts £64,862 (excluding £8,000 already written off as bad debt), it is surprising that they had any intention of continuing to trade with them. The Jowitt Board allowed Scotts to continue selling to Fraser & Wilson only on the basis that they paid off £1,500 of their debt to the bank and £1,500 to Jowitts every month and paid on monthly terms for new purchases – that is, if they wished to buy £1,000-worth of cloth in a month, they would have to pay £4,000.

The Scotts informed the Jowitt Board that they were making drastic cuts to their mill expenses, dismissing many hands and placing all their foremen on time wages. They intended to reduce the scope of their trading and only accept prompt-paying customers. They would cease the fancy worsted trade and deal only in plain worsteds and fancy woollens. However, they asked Jowitts for permission to obtain outside finance for the purchase of black-face wools and splash yarns.[406] The Jowitt directors were unable to agree to this, although they were willing for Scotts to use part of the £30,000 facility for such purchases. However, the purchases would have to go through Jowitts who would charge 4% commission. Further, they did not wish to put the new agreement in writing, although 'it was their intention' to grant credit up to £30,000, subject to the conduct and position of Scotts and the continuance of the banking arrangements.[407]

Although realisation of old stock and book debts was believed to have reduced Jowitts' losses on the joint venture substantially by September 1923,[408] the company was still facing a ubstantial deficit. The liquidation of Fraser & Wilson was very slow and Jowitts were still urging that it should be wound up without delay in January 1924. At the same time, the Board asked Gerald Veale to write to the Commercial Bank of Scotland, agreeing to the payment of all creditors other than the bank and Jowitts.[409] Later that month, Jowitts agreed to accept a dividend of five shillings in the pound on its claim against Scotts and on the joint account, 'provided the 5/- is guaranteed by the Bank.' Jowitts also asked the Commercial Bank of Scotland to release it from its £5,000 guarantee, and emphasised that all the assets in the joint account belonged to Jowitts. George Blackwell was authorised to act for the company in the liquidation of Walter Sott & Sons Ltd.[410]

In July 1924, the company's total loss on the joint venture was calculated at about £100,000. Walter Scott & Sons Ltd ceased to exist but Fraser & Wilson survived in some form. Presumably because they could not find a buyer at the right price, Robt. Jowitt & Sons bought Fraser & Wilson's property in Queen Street, Dumfries,[411] which they retained until 1927, renting it back to F&W. Jowitts also made a loan to Fraser & Wilson, complaining in 1927 that the company's accounts were unsatisfactory.[412] The £1,700 still outstanding in 1931 was finally settled for £1,500 cash.[413] Two Scottish companies had been largely destroyed and Jowitts had lost a fortune. So came to an end the inglorious experiment known to the Scottish press as 'the woollen trade fusion'.

ARCOS LTD – WOOL AND ESPIONAGE

With a stagnating economy in western Europe, Russia presented a potentially large market for Jowitts, but one where the still uncertain political and economic environment presented equally large risks. With serious concerns over payment, Jowitts used Kleinworts to discount Russian bills.[414] The other problem was the shifting nature of the Russian commercial apparatus. At the beginning, much of Jowitts' Russian trade was with the All-Union Textile Syndicate (AUTS) through the agency of Paul Ergounoff. In July 1927, contracts with the AUTS totalled £71,437.[415] Jowitts also signed an agreement of an unspecified nature with Ergounoff and Prince Sergei Obolensky.[416] Ergounoff spent much of his time abroad, possibly including the Soviet Union, but Obolensky's role is more puzzling. As a leading White Russian, he was *persona non grata* in the Soviet Union and it is doubtful whether he would would have been an acceptable intermediary between Jowitts and any Russian trade organisation. By this time, Soviet wool-buying in the West was already being centralised under the All Russian Co-operative Society (Arcos Ltd). This Russian-owned trading company was registered in England with branches throughout the British Empire and sister organisations in other parts of the world. Arcos became a significant customer for Robt. Jowitt & Sons, especially for their Cape wools. In 1924, after protracted negotiations, Jowitts offered Arcos up to £10,000 credit and a rebate of 1% on turnover with them of up to £250,000 per annum, with a minimum of £100,000.[417] In the following year, the Board resolved to sell tops to Arcos up to a value of £12,000 per month, accepting 40% cash, 30% bank acceptances and 30% Russian bills.[418] Paul Ergounoff seems to have played a minor role in dealings with Arcos but his services were dispensed with in May 1929. His advances on commission were never covered by his earnings and were written off as bad debt.[419]

Wool was only a small part of Arcos's trading activity and, in fact, trading was largely peripheral to its true purpose; its senior managers were mostly handpicked officers of Soviet Military Intelligence. Since the Government Code & Cypher School had broken the Soviet military and intelligence codes as early as 1920,[420] the British Government was well aware of the company's true role as a front organisation for Soviet espionage and subversion. Indeed, MI5 had an informant within Arcos who supplied them with photographs of documents between 1924 and 1927.[421] On 12 May 1927, police and MI5 officers raided 49 Moorgate, the home of Arcos Ltd and the Soviet Trade Delegation and found two men and a woman burning incriminating documents in the basement. The raid was described at the Communist International on 18 May as 'a dastardly attack on the liberties of the proletariat.'[422] On 24 May 1927, the Prime Minister, Stanley Baldwin told the House of Commons:

In conclusion, it may be pointed out that the evidence now in the hands of the authorities proves that:

1. Both military espionage and subversive activities throughout the British Empire and North and South America were directed and carried out from Soviet House.

2. No effective differentiation of rooms or duties was observed as between the members of the Trade Delegation and the employés of Arcos, and both these organisations have been involved in anti-British espionage and propaganda.[423]

The minutes of a Cabinet meeting on 23 May recorded that,

... there was general agreement that the Russian Trade Agreement should be terminated and that the Russian Trade Delegation and all individuals in Arcos known to be engaged in propaganda work should be expelled.

However, it is a reflection of the Soviet Union's growing economic importance to the United Kingdom that they went on to say,

It was not considered desirable, however, that the trading activities of Arcos (a British company whose capital is owned by the Soviet Government) should be brought to an end by British Government action, as it was desirable that trade with Russia should continue on the same basis as it is conducted in the United States of America and other countries which have no Trade Agreement with the Soviet.[424]

Thus it was that a known Soviet front organisation continued to trade in the United Kingdom and elsewhere in the British Empire. It is also clear that, although much of the Russians' later business was legitimate, their espionage activities did not cease in 1927. For instance, Hugo Rudolf, an Arcos employee in Egypt, was identified as a Russian spy in 1929,[425] Michael Gansovitch Sheffer, an Arcos employee in London doubled as an OGPU agent,[426] and another Arcos employee, Emily Wooley was a Comintern courier during the thirties.[427] Naturally, companies in Britain and elsewhere became rather wary of dealing with Arcos for a

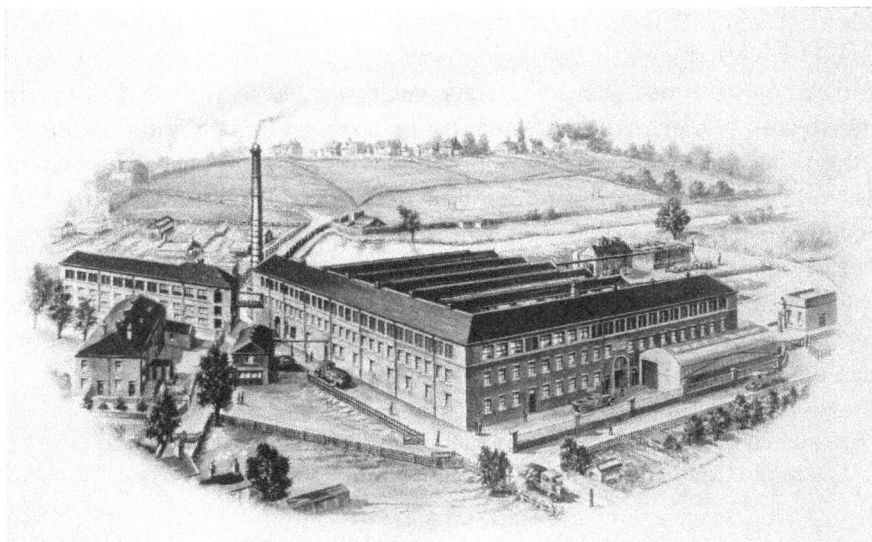

Fig 36 The Fellmongery at Meanwood, Leeds, in about 1925.

Fig. 37 The Combing Shed, Bradford, in about 1925.

while. When J. Patiagorsky of the Textile Import Company Ltd of Moscow visited Sydney in 1927, he felt compelled to state that his company and Arcos were not allied.[428] Arcos was dissolved in 1927 and the staff transferred to Germany but, since trade between Britain and the Soviet Union was important to both parties, it was reconstituted two years later. The new Arcos bounced back, moving its offices to Bush House, now the home of the BBC World Service, in 1929.

Robt. Jowitt & Sons was itself the victim of espionage in relation to its Russian trade. George Blackwell reported to the Board on 19 July 1929 regarding the leakage of information, in particular that relating to Russian contracts, assuring the Directors that,

> steps had been taken to prevent such information being available to anyone in the Company's employ, other than those directly concerned.[429]

Arcos achieved a measure of respectability in 1934 when the Russian government sold the major portion of the *Codex Sinaiticus* through Maggs Brothers to finance Stalin's second Five Year Plan. Bizarrely, it was Arcos which delivered the codex (wrapped in cotton wool in a brown paper parcel)[430] to the British Museum and received the cheque for £100,000.[431]

While rejecting a request for nine months' credit,[432] Robt. Jowitt & Sons traded heavily with the reconstituted Arcos and even diversified into selling camel hair.[433] The company drew up a new agreement with Arcos in 1933, insisting on 30% up-front in cash and the rest in four-month bills.[434] Arcos proved to be demanding customers (and, indeed, was involved in many court cases with suppliers throughout the 1920s and 1930s, both as plaintiff and defendant). On a shipment of 20 tons of scoured wool, they demanded a reduction of 2d per lb (a total of £373 6s 8d) from Jowitts. On arbitration, this was reduced to ½ d per lb on 33 bales only.[435]

AN UNEASY ALLIANCE

Although there is never a hint of it in the minutes of Board meetings, there was, almost from the beginning, a great deal of tension between the younger generations of Jowitts and Harlands which persisted until the Harlands' involvement in the company ceased in the mid-1960s. In

Fig. 38 The combing shed, Hollings Mill, in about 1925 showing Noble combs and finishing boxes.

essence, the Jowitts considered the Harlands rather stupid while the Harlands considered the Jowitts idle and privileged. Somehow, despite this conflict, or perhaps because of it, the company usually functioned rather well. A Jowitt was always Chairman and, until Peter J. M. Bell's appointment in 1967, a Harland was always Managing Director.

Most of the members of both families managed to rub along with the opposing camp, the Harlands receiving not-inconsiderable pay cheques for their efforts. One Jowitt who could not rub along was E. M. Jowitt's son, Anthony Thomas McCulloch Jowitt – often, like his father, known as Tom. He could not abide the Harlands and decided in the mid-twenties that the silver screen rather than wool was his true vocation. A leaving party was arranged for him at the Great Eastern Hotel, although it is doubtful if anyone, even Tom, thought that he would succeed.

He proved them wrong. Almost as soon as Tom (who reverted to his first name, Anthony, for his film career) arrived in Hollywood, he started appearing in films – often rather raunchy ones – opposite the most al-luring film actresses of the day. In 1925 alone he appeared in *The Little French Girl, The Lucky Devil,* with Esther Ralston (the 'American Ve-nus') and Edna May Oliver, *The Coast of Folly* with Gloria Swanson, *A Woman of the World* (only a bit part) with Pola Negri and *The Splendid Crime* with Bebe Daniels. In truth 1925 was the high-point of his career, although he appeared in a number of other films up to 1950. He had dif-ficulty with the transition to the talkies because, unlike the many theatre actors who had found themselves in Hollywood, he had only learned to act in silent films. In 1932 he appeared as Clara Bow's boyfriend, Jay Ran-dall, in *Call Her Savage,* a bizarre film which includes rape, prostitution and a gay bar. In that year he also had a bit part in *Mata Hari,* with Greta Garbo, but for the most part his Hollywood career consisted of working as a dialogue coach. At least he never had to return to Bradford and the Harlands. Anthony also made a number of films for the US government. His daughter, Deborah, became an immensely successful dancer, chore-ographer and long-serving dance critic for the *Village Voice.*

RETURN TO THE GOLD STANDARD

When the Chancellor of the Exchequer, Winston Churchill, on the ad-vice of Montagu Norman, the Governor of the Bank of England,[436] re-introduced the Gold Standard in 1925 (with the pre-war parity of $4.86 to the pound), virtually all British products became uncompetitive. This was felt particularly in the coal mining industry, which also suffered from increased competition from producers in the United States and Poland, not to mention 'free' coal exported by Germany to France and Italy as part of the reparations. The response of British mine owners was to force down wages, triggering the General Strike of 1926.

Other British industries, including wool, were seriously affected by the return to the Gold Standard. In fact, the impact of the rise in sterling and widespread labour disputes on Robt. Jowitt & Sons was brutal, with the company plummeting from a profit of £56,700 in 1924 to a loss of £67,400 in 1925 (see Appendix IX). The following years were extremely volatile. Wool prices rose in 1927 as a result of a drought in Australia, and then slumped back in 1928. The company made modest profits in 1926 and 1927 and only £4,800 in 1928. The British economy had not fully re-covered from this economic disadvantage when the Wall Street Crash of

Fig. 39 Anthony Jowitt who escaped the wool industry
and the Harland family for a film career in Hollywood.
The photograph, taken about 1930, is reproduced
courtesy of Paramount Pictures.

29 October 1929 caused international financial panic and ushered in the Great Depression. It is hardly surprising that Robt. Jowitt & Sons Ltd had a run of losses in 1929, 1930 and 1931 (Appendix IX).

In 1929, Robt. Jowitt & Sons, along with other employers, responded to the extremely difficult trading conditions by attempting to reduce wages by 10%, provoking calls for strike action in many sectors of the industry. Union leaders advised their members to avoid strike action, but many of their members were not in a mood to listen. Rather than an all-out strike, a series of piecemeal actions affected only some companies at a time. Despite their initial reluctance, the union leaders supported their members, even in some cases paying strike pay to non-members to make the action more effective.[437] Starting on 9 November 1929 with industrial action by the overlookers, Jowitts and all other textile companies were hit by a protracted series of strikes and it was not until July 1930 that the combing operatives were back at work. Jowitt's existing mill bonus scheme, based on combing output, ensured that combing operatives had every incentive to avoid industrial action and, in general, the company's workers were usually the last to strike. If, as in this case of concerted industrial action on a national scale, a strike was unavoidable, they were equally incentivised to make up for lost production as quickly as possible. Indeed, in their first month back, workers at Hollings Mill worked so hard that they received a 51% bonus. However, the company found it impossible to reach an agreement with the wool-sorters. Provoked to uncharacteristic ruthlessness, the management replaced nearly all the pre-strike sorters and foremen. In July 1930, the replacement workers were reported to be working a forty-eight-hour week 'without stint'.[438]

Britain left the Gold Standard on 19 September 1931, which was generally beneficial, although it left the company with the problem of paying higher sterling prices for forward contracts in foreign currencies. Australia had already left the Gold Standard. South Africa, whether through misguided nationalism or financial ignorance, clung on to the Gold Standard for some time, making its produce, including wool, very expensive to British buyers. It eventually came off the Gold Standard on 27 December 1933.

MORE TROUBLE AT THE LONDON OFFICE

The London Office, which had been such a problem in the early years of the century, continued to cause trouble and, in 1926, H. C. (Cecil) Edwards, who had succeeded his father F. M. Edwards of rabbits and butter notoriety, was himself dismissed. Like his father, he owed the company money, but this debt was forgiven and he was given a cheque for £200. There was clearly a problem with the financial arrangements for managers, since they were so often indebted to the company. However, Edwards was receiving the substantial salary of £400 per annum in 1920. The general problem seems to have been that managers were allowed to borrow against profit-sharing bonuses which did not always materialise. Edwards was succeeded by George F. H. Eke who also had financial problems. In 1932, A. T. Keeling spoke to Mr Eke about it and it was agreed that he could borrow £60 from his building society, secured against his house, the building society's claim being given priority over the company's. Mr Eke's assistant, B. Priggen, was summarily dismissed in April 1932, following a report on his conduct from Mr Eke. At the same time, following a report from the accountants, Mrs Margaret Baines was given three months' notice. In the light of her long service, she was given a cheque for £100 and forgiven debts of nearly £23.[439] Despite the involvement of the accountants, gossip in the office had it that their transgressions were of a sexual rather than a financial nature. The Jowitts might no longer be Quakers, but they remained a little straight-laced.

George Eke remained indebted to the company for several years and, in October 1934, he informed the company of his intention to sell his house. He proposed to pay the company £50 from the proceeds, repaying the remaining balance of approximately £68 by six shillings per week, which suggestion was accepted by the Board.

THE DOWNFALL OF THE COMPANY SECRETARY

Everyone liked George Blackwell, the Company Secretary; they admired his fine house at Pool Bank near Harrogate, they liked his charming manners and, of course they trusted him completely. At the beginning of the thirties, the company was even more vulnerable than usual to sudden fluctuations in wool prices and Blackwell tried to persuade the directors that they should hedge against potential losses by using wool futures. These were not then traded in London, although they were traded in

Antwerp and on the Roubaix Wool Exchange in France. Despite the proposal's obvious merits, it was rejected as too radical. The directors' hesitation was understandable; those with long memories could recall the bubble on the Roubaix and Antwerp exchanges in 1900 which caused a number of companies involved to suspend payments, and at least seventeen wool companies to fail. As one commentator remarked,

> There is not the least shadow of doubt in the minds of anybody conversant with the gambling operations at Antwerp and Roubaix that prices have been forced down in a most determined manner by the 'bear' tactics which have been brought into play for pure speculative purposes, and which neither the exigencies of the trade nor supplies either warrant or suggest...

> Towards the close of last week news reached Bradford of the failure of these important Roubaix firms with liabilities amounting to 15,000,000 fr, or about £552,000. It need hardly be said that the news in Bradford created quite a sensation, and especially so when the number increased with wonderful rapidity, no less in aggregate than 17 firms being compelled to consult their creditors. The amount of liabilities is estimated at £3,500,000 though it must be said that there is every likelihood of the total reaching not far short of four millions sterling...[440]

The same writer claimed that the whole Australian wool clip had been sold forward several times over.

Unfortunately, George Blackwell was only about thirteen when the wool futures collapsed in 1900, too young, it seems, to have learned from it. He decided to put his plan into effect without the directors' consent, gambling on futures on the Roubaix exchange through the company's agent, Jules Florin. Sadly, he lost money on the deals – and not just his own money. As had been the custom in partnership days, Directors and some managers were in the habit of keeping some of their money in accounts with the company rather than with banks, and it was found that Blackwell had made unauthorised withdrawals from these personal accounts. The sense of betrayal which must have been felt by the members of the Board is hard to over-estimate. Not only had George Blackwell been the Company Secretary since the company's incorporation and a Director, he had also been one of the trustees appointed under F. McC. Jowitt's will, a position of the utmost trust. At the board meeting on 28 December 1931, the company's solicitor, A. T. Keeling, reported to the

Fig. 40 George Blackwell, the much-liked Company Secretary, before his ignominious fall.

board that he had spoken to George Blackwell who had confirmed that he was insolvent. Keeling informed Blackwell that under article 83 of the articles of association his position as director was now vacated. The minutes note that,

> On receipt of this Report the Directors, with very much regret decided that Mr. Blackwell's connection with the Company be and is hereby determined...[441]

This was, no doubt, how the situation was represented to the outside world. It was embarrassing enough that the company secretary should have to stand down because he was insolvent. The minutes of the AGM held on 19 February 1932 stated simply, 'Mr. G. Blackwell has ceased to be a Director of the Company.' However, the true state of affairs was made clear in the remainder of the Board minutes:

> ... in consequence of his personal account with the company, and of irregularities in the state of certain private accounts of members of the Company

147

and staff, with the Company, which accounts were under the control of Mr. Blackwell. Mr Blackwell's authority for signing any document of whatever kind on behalf of the Company was withdrawn.[442]

He was immediately replaced by Mr Charles Scott Tullie who had been Assistant Company Secretary since 1924. With a generosity which seems to have been characteristic of the company's dealings with miscreant employees, Keeling informed Blackwell that it had been decided to pay him a weekly allowance 'to be entirely at the discretion of the Board'.[443] The company continued to pay George Blackwell £5 per week up until the end of May 1932, and even then extended this payment for another three weeks.[444]

The company had insured Blackwell's life, as they had the lives of other senior staff. At the board meeting on 1 July 1932, the board discussed the surrender of the policy (which was fully paid-up) but decided to leave it in force for the time being.[445] In the event, he lived until the end of 1960, when the policy yielded the company £22,728.[446]

The Company Secretary's downfall prompted a general tightening of financial controls. In particular, Card Memorandum Accounts, whereby the company effectively acted as bankers to directors and others, were phased out in favour of monthly payment of salaries by banker's order. It was also resolved that,

> it be an instruction to the cashier that no 'subs' be made to an employee on account of his or her salary or wages, without the express instructions in writing of the Managing Director.[447]

At the same time as George Blackwell was dismissed, his younger brother, Joshua Millington Dixon Blackwell, the Assistant Manager of the Sydney Branch, was given notice to quit his job. This was probably unrelated to George's accounting irregularities as Joshua was substantially in debt to the company, presumably due to advances on salary or commission.[448] He had also been named as co-respondent in the divorce of Herbert Drysdale Varley and his wife Rose (who later, as Rose Skinner, became an art dealer famous for promoting Australian artists such as Sidney Nolan). Drysdale demanded damages of £5,000 from Joshua Blackwell, but later settled for £300.[449] It was still a substantial sum for a wool-buyer to find and may explain his indebtedness to Robt. Jowitt &

Sons. The termination of his employment was said to be 'under the provision of clause 7 of Mr. J. M. D. Blackwell's loan agreement,'[450] but there must also be a suspicion that the company was unhappy to have been associated with a high-profile divorce case, something which might have jeopardised his career earlier if he had not been the Company Secretary's brother. In a judgement which seems harsher than that on his brother, the minutes record,

> It was further reported that it was not anticipated any sum would be recovered in respect of Mr. J. M. D. Blackwell's indebtedness to the Company and that for the purpose of the Company's Accounts, the amount should be written off, but without prejudice to the Company's Right to Action.[451]

It was left to the discretion of Arthur Gibson, Manager of the whole Australian operation, to decide whether Joshua Blackwell's fare home would be paid by the company. It was reported in the minutes for 29 June 1932 that his fare had been paid and that he had sailed for England. However, his return to Britain was only brief, as the census records show that he was still working as a wool buyer in Sydney in 1937. He died in Newtown, NSW, in 1941.

Another Australian employee, Jack Burgess in Melbourne, was overdrawn with the company and he was dismissed in 1932.[452]

THE CONTINUING RECESSION OF THE 1930S

The world was slow to recover from the Great Depression. In Britain and continental Europe, company failures, bad debts and extended credit were a constant problem (although, fortunately, much of the foreign debt was insured by the Export Credits Guarantee Department). Recognising the seriousness of the situation, the Directors decided in September 1931 to reduce their own salaries (except those of George Blackwell, A. A. Gibson and D. B. Sykes) by 20% and those of all other staff by 10% from 20 November 1931.[453] The decision with regard to non-directors was rescinded[454] but the salaries and fees for Sam Harland, E. M. Jowitt, A. T. Keeling, W. T. B. Jowitt, R. B. Jowitt II and G. C. Veale were duly reduced by 20%. Robert Jowitt II had announced in September 1931 that he intended to take a six month leave of absence and, at his suggestion, his salary was reduced by 50% for this period.[455]

The company was exposed to serious outstanding debts in Germany, Greece, France, Poland, Portugal and Italy, and actually suspended all trade with Germany for a time. As a result, much of the board's time was taken up considering and revising customers' credit limits. In 1932, its French agent, Jules Florin, pressed the company to extend its credit terms to 90 days, but Sam Harland told the board that he had established that such terms were unusual and the board declined the request.[456] By 1935, the German debts had almost been cleared, to the point that their financial position was described as 'very satisfactory'.[457]

THE WOOLCOMBERS' MUTUAL ASSOCIATION

Robt. Jowitt & Sons, like the other large wool companies, had a combing mill which combed wool for its own Top Department but used comission combers for over 50% of its top production. Unfortunately, due to falling demand, there was serious over-capacity in the British combing industry from the 1920s onwards. This over-capacity led to falling combing prices, which, in turn, made investment in new machinery unattractive, leading to less efficient production and even tighter margins. Moreover, they were facing competition from smaller companies whose machinery was even older and had therefore been almost completely written off.

Like a number of industries, the woolcombers were facing complete devastation by the early 1930s and it was clear that they must cut capacity to survive. The Woolcombing Employers' Federation (WEF) started seeking a solution to this problem in 1930. They turned for their model to the National Shipbuilders' Security Limited (NSSL), which had been formed in that year with the aid of the Bankers' Industrial Development Company. Financed through a 1% levy of the sales of participating companies, the NSSL aimed to 'assist the Shipbuilding Industry by the purchase of redundant and/or obsolete shipyards, the dismantling and disposal of their contents and the resale of sites under restriction against further use for shipbuilding'. The WEF proposed a Woolcombers' Mutual Association (WMA) comprising its topmaking and commission combing members.

The Harland family had joined Jowitts when that company had entered the woolcombing business with the purchase in 1900 of S. & S. Musgrove. So, it was particularly appropriate that Sam Harland, Jowitts' Managing Director, was their representative in the discussions which led to the founding of the WMA. He reported to the board regularly on its progress and by September 1932 was able to give them details of the draft contract.[458]

The Woolcombers' Mutual Association was incorporated on 21 February 1933 and was authorised to trade on 20 March of that year, with Sam Harland as one of the founding directors. At the same time, the Commission Woolcombers' Association (CWA) was created. All commission woolcombers in the WMA were also members of the CWA. The other three groups in the CWA were the topmaker combers, who owned the wool and the combing plant in which they produced the tops, the spinner combers, and the hair combers. The CWA and WMA entered into an agreement regulating the prices for combing. The WMA's rules were strict, as they had to be if they were to achieve their aim. For wool not combed in their own plants, the topmaking combers in the WMA were only allowed to place wool for combing with members of the CWA – at, of course, the regulated rates. Equally, topmaker comber members of the WMA would charge the agreed CWA rates for any commission combing which they undertook. If a WMA member regularly accepted commission combing, it would be obliged to join the CWA.

The initial allotment of 9,000 £1 shares in the WMA was made in direct proportion to the subscribers' top production, and it was a rule that shares in the Association held by a company which ceased to comb were to be sold the remaining members on a pro rata basis. Robt. Jowitt & Sons received 162 shares in the initial allocation, reflecting their relatively small combing capacity. They also applied for £1,500-worth of debentures.[459] The WMA wasted no time in putting their capacity reduction plans into effect, and by May 1933, they had bought Valley Woolcombers in order to take it out of production.[460] By the end of 1933 they had bought a total of twelve concerns (although three were from the WEF, which had itself bought them to take them out of production).[461] It was the WMA's official position from the beginning that all machinery bought by them should be destroyed. While this suited the WMA's intention to reduce combing capacity, it was an expensive business, paying the full commercial price for machinery which had only scrap value. By 1938, when the WMA bought up Holden, Burnley's combing plant, they had almost exhausted their funds. At the same time, it seems to have been almost too successful because, by then, there were some concerns in the industry that there was *under*capacity.[462]

At least equally important as the scrappage scheme was the regulation of combing charges and since the WMA membership accounted for 89% of all combs and 91% of all topmaking in the UK, they had almost total control of prices in the market.[463] It is unlikely that such arrangements would be allowed under modern competition law, but at the time they were both necessary and legal. The success of the scheme was indicated by the fairly substantial rises in tariffs between 1931 and 1937 in difficult market conditions.[464]

A NEW MANAGING DIRECTOR

On 18 January 1935, the board appointed George Harland as assistant to his father, the Managing Director, 'to hold authority in his absence to carry out the board's policy.'[465] Also in early 1935, Sam Harland, who had been Managing Director since the company was incorporated in 1919, informed the Chairman that he would like to take things more easily 'at no distant date'. E. M. Jowitt proposed that when Sam Harland chose to step down, he should be replaced by George Harland and FrederickMcC. Jowitt's son, William Thomas Benson Jowitt, as joint Managing Directors. Sam Harland would act in the capacity of Advisory Managing Director. George Harland was made an Ordinary Director in April 1935, but his father did not step down as Managing Director until the end of 1939,[466] when he was about sixty-five.[467] At this point, E. M. Jowitt also stood down as Chairman, his place being taken by W. T. B. Jowitt, the idea of joint Managing Directors being dropped. At the board meeting on 12 January 1940, Maurice and Robert Jowitt II expressed their appreciation to Sam Harland for his 'very successful Managing-Directorship'.[468] Sam Harland's salary was reduced to £1,500 p.a., plus 5% commission on the profits, George Harland's salary was increased to £2,250 and E. M. Jowitt's to £1,750. In the latter two cases, it was stated that, 'bonuses, (if any,) to be fixed by, and at the discretion of the Board.'[469]

Sam Harland was not the only person who felt the need to take things more easily. The chief London wool-buyer, Joe Sheard (whom the Board had considered putting on the pensions list in 1932),[470] retired on 20 November 1938 after sixty-two years with the company. He had started at the company's Albion Street offices at the age of 14½ in 1876, when John junior was still actively involved. The company gave him a gift of £100 and an pension of £200 p.a. to be reviewed annually. This pension, like all others given by the company at the time, was termed a 'voluntary allowance' and was at the discretion of the Board.[471] However, there seem to have been no cases in which the Board sought to reduce or cancel such an arrangement during the recipient's lifetime. Sadly, only seven months later it was reported to the Board that Mr Sheard was totally blind, although this might be relieved by an operation costing approximately sixty guineas. The Board voted to make a grant of this amount to Mr Sheard and it was later reported that his sight was recovering.[472] In July 1940, the company learnt that Mr Sheard had financial problems and lent him the money to pay off his bank overdraft at 4½% interest p.a., against the security of his life policy, on which the company undertook to pay the premiums. The terms do not seem to have been over-generous for such a long-serving employee, especially as the company had £75,000 of 'surplus finance' at the time.[473] Mr Sheard hardly lived long enough to benefit from this arrangement as he died at his home at Stanmore Crescent, Leeds, on 9 August 1940, aged seventy-eight.[474]

A FRENCH ADVENTURE

The company became involved in fellmongering on commission and trading sheepskins in the thirties, maintaining a rented warehouse at Mazamet in the South of France. Alfred Salvisberg became their agent there in 1935. There were a number of business failures in the French wool trade in 1937, attributed to a fall in the franc and weak economic conditions. Jean Roux & Cie., one of their debtors, seemed to be in trouble and George Harland and the Secretary, Mr Tullie, had visited France to assess the situation. Tullie reported to the board that the amount outstanding was £16,497.[475] An investigation by George Harland and a Monsieur Montmain from Peat, Marwick, Mitchell's Paris office established that Jean Roux & Cie, was 'hopelessly insolvent'. Fortunately, Roux had pledged a substantial amount of wool, deposited with the bourse in a public warehouse, as security for their debts, and Jowitts were assured that this was enforceable in French law. The amount secured by the wool was unlimited, and guarantee applied to all transactions up to the date at which their agreement had been terminated, which was 27 May 1937. The company, fearing prolonged French legal proceedings, proposed to obtain this wool from the warehouse and hand over £400 of assets (later increased to about £700) to other creditors of Roux.[476] All parties agreed and the French court gave its approval. Moreover, a Mme Valat, widow of Armand Valat, had guaranteed those of Roux's debts which were not covered by the wool.[477] In 1938, Madame Valat asked for renewal of a bill which was about to mature and offered to pledge wool against it. George Harland visited Mazamet again and, after investigation, obtained a court order for restraint of goods and bank account, the property to be subject to the decision of the court.[478] This action seems out of keeping with the traditional ethos of the company, but it must be remembered that Robt. Jowitt & Sons had lost more than £64,000 in 1937 (Appendix X) and the management had a duty to protect shareholders' interests. Mme Valat's property included a farm, mill, adjoining house and land. Another creditor, a Madame L'Estrade also had a claim on some of the property and agreement was reached that the land would be sold, with all the proceeds going to Robt. Jowitt & Sons, together with 50,000 francs from Mme Valat's bank account. The balance of the debt would be covered by a joint mortgage with Mme L'Estrade on the mill and house.[479] M. Salvisberg advised that it might be better for the mortgage to be in the name of an individual and it was agreed that it should be in George Harland's.[480] The house was quickly sold for £1,830 5s 0d nett and the farm soon after for 80,000 francs.[481]

By October 1938, there was still 555,770.15 francs debt outstanding and it was proposed that Robt. Jowitt & Sons and Mme L'Estrade be given an option to purchase the works for a fixed price from Mme Valat's estate upon her death.[482] In May 1939, Mme Valat's relatives suggested that they take over the company's share of the mortgage to the amount of 237,000 francs (then worth £1,339). M. Salvisberg informed the board that there was a prospective market for the mill at 600,000 francs, the company's share of which, together with a cottage, would amount to approximately £1,600. The board resolved to try to achieve this figure, but to give preference to Mme Valat's family at an equivalent price.[483] By mid-June 1939, the matter was successfully concluded, with the company receiving £1,843 15s 3d, less fees.[484] By September of that year, the world was at war.

The company continued to rent the warehouse in Mazamet, renewing the lease after the war at a rent of 120,000 francs per annum,[485] only about £122 at the prevailing exchange rate.[486]

World War II

My good friends, this is the second time in our history that there has come
back from Germany to Downing Street peace with honour. I believe it is
peace for our time. We thank you from the bottom of our hearts. And now
I recommend you to go home and sleep quietly in your beds.

Neville Chamberlain, 30 September 1938[487]

PREPARATIONS FOR WAR

Despite Neville Chamberlain's foolish optimism, war never looked very
far away in the nineteen-thirties. Fascist dictatorships had come to power in Italy and Germany, and in 1936 Hitler had sent the Condor Legion
to fight for fascism in Spain. For this reason, it is no great surprise to find
the board of Robt. Jowitt & Sons discussing air-raid precautions as early
as 1936.[488]

By then, even the British government was aware of the danger and had
started re-arming. On 20 April 1937 Neville Chamberlain, then Chancellor of the Exchequer, announced a new tax to pay for the increased costs
of this re-armament, to be called the National Defence Contribution. It
was intended to tax companies who benefitted from the preparations for
war, but it was unclear which companies it would affect or how much it
would cost them. Robt. Jowitt & Sons awaited clarification,[489] presumably fearing that the supply of wool for uniforms would be included. In
the event, the political opposition to the tax from Chamberlain's fellow
Conservatives was so strong that he had to drop it on 1 June, just five days
after becoming Prime Minister.[490]

At this time, the Government also started recruiting more troops
for the Territorial Army. In 1938 it was agreed by the company that two
weeks' paid leave would be given to staff who signed up, to allow them
to attend training camps.[491] Also in that year the government improved
air-raid precautions, and four of the company's staff attended one-week
ARP anti-gas courses.[492]

On 5 May 1939, the board agreed that employees and directors who
were employed by the company at the outbreak of war and then served
in the armed forces would have their service wages augmented by the
company,

... such as to bring the employee's earnings to the same wage or salary as that received during active employment with the company.[493]

However, by October, no doubt influenced by the realisation that the war could go on for several years, this had been altered to the company paying them

three-quarters of the difference between their service pay and allowances and what they would have received if they had remained in the employ of the company, but subject to the right of the company to investigate each case, and make such adjustments of the foregoing basis as meets with the approval of the Managing Director, having regard to the individual circumstances of each particular case, and having regard also to the financial capacity of the company to continue making such payments, the company reserving the right to review the whole matter from time to time.[494]

It was, nonetheless, a very generous gesture on the part of the company. The rules were altered several times over the war years to include more staff. In September 1941, a Mr J. E. Howdle, who had not joined the company until five days after the outbreak of war was granted the allowance.[495] Apprentices who joined the armed services and had reached the age of twenty-one were considered in 1942 and it was agreed that they would be deemed to have been receiving a qualified employee's day-rate for the purposes of the scheme.[496] The allowances cost the company a total of about £20,000.[497]

ENEMY DEBTS

Although the company had made air-raid precautions as early as 1936, it had continued trading with Germany until the eve of war and still had several thousand pounds outstanding in September 1939. Unsurprisingly, the German debt was still outstanding in January 1940 when C. S. Tullie informed the Board that it amounted to £4,347.[498] Fortunately, this trading had been insured with the Export Credits Guarantee Department, and the debt was eventually settled.

Few in thirties Britain had anticipated that the country would soon be at war with Japan, so it is not surprising that Robt. Jowitt & Sons was caught out by this event, with 200 bales waiting in Durban for shipment to a Japanese customer. The wool had cost £3,448 8s od and had attracted charges of £381 2s 11d by August 1941. The bales were sold to the British Wool Commission at a loss of £240 in November of that year.[499]

In July 1941, after Finland had allied herself with Nazi Germany, Finnish debt stood at £3,102 6s 7d. This was later classified by the Board of Trade as enemy debt and was therefore covered by the Export Credits Guarantee Department.[500]

WOOL CONTROL

Just as the British government had taken control of wool stocks during the First World War, they did so again in the Second. Indeed, the government had started taking measures at the time of the Munich Crisis of 1938, and a shadow organisation for wool control was in operation well before war broke out. On 1 September 1939, all dealing in privately-owned wool was prohibited, but trading was restarted on 3 September, under licence. The Ministry of Supply took control of 'wool, tops, broken tops, noils, and combing laps,' with this authority vested in the Wool Control. Sir Harry Shackleton, President of the Wool Textile Delegation, was made Wool Controller. Also in September, the British government agreed to purchase the total Australian and New Zealand wool clips, not just for the duration of the war, but for one year after as well, at an agreed fixed price. Naturally, it also purchased the total UK wool clip. On 30 October 1939, the Wool Control acquired all UK wool stocks not in the hands of spinners and manufacturers, at scheduled prices. In early 1940, the Wool Control also managed the collection and disposal of UK sheepskins. Later that year the British government purchased the whole South African clip on similar terms to Australia and New Zealand.[501] Many of Jowitt's Australian staff became appraisers for the Australian Wool Control, and at higher salaries than those paid by Jowitts, with the exception of Mr Dudley V. Yeoward, their manager in South Australia, whose lower salary as an assessor was made up to his previous pay by the company. Similarly, two British employees who were transferred to the Wool Control had their salaries made up by the company.[502] At the company's request, Mr George Rushby relinquished his role as a Preliminary Appraiser at £A900 p.a. in order to supervise the Adelaide fellmongery business. The company accepted Mr George Rushby's suggestion that he should forego all remuneration from Jowitts while working for the Australian Wool Control, but said that he should still receive the bonuses based on

Australian sales due to him under the agreement expiring 20 November 1939. Mr D. B. Sykes and Mr W. R. Myers of the company's South African office were employed as Chief Appraisers in the British Wool Commission and also received higher salaries than they had with Robt. Jowitt & Sons.[503] By December 1940, Mr Morrison in their Durban office was the only member of the South African staff who had not been absorbed into the Wool Commission.[504] Mr Keeling was requested to draw up a waiver of directors' fees for the duration of the Control period.[505]

The wool companies were required to provide details of all their wool usage and all public contracts they already held. The Wool Control then distributed wool for processing *pro-rata* to previous activity. Sam Harland informed the Board that the Woollen Department were acting as purchasing agents on a 1½% commission and that a similar arrangement was anticipated for tops.[506]

It is clear that the careful planning which had been put into the Wool Control saved many companies, including Robt. Jowitt & Sons, from complete disaster. Although much of its commercial activity was curtailed by the government monopoly in wool, its wages bill was substantially reduced and the expertise of its employees put to good use by the Wool Control. Meanwhile, it was still able to make money and provide employment by working on commission for the government. In fact, the company made no losses (though only modest profits) during the war.

THE CONCENTRATION SCHEME

Since the outbreak of war, the British Government had curtailed the production of civilian supplies by the use of Limitation of Supplies Orders, but in 1941 it was decided that even these were not drastic enough to redirect labour to essential war work. On 6 March, Oliver Lyttleton, President of the Board of Trade, proposed an industrial concentration scheme whereby companies would be identified as either 'nucleus' firms or 'satellite' firms, the former to be worked at full capacity and the latter to be closed down for the duration of the war. While the Government indicated the amount of capacity which should be removed from each sector, it left it to individual companies to formulate the concentration schemes. Nucleus firms had access to raw materials and their employees were allowed to defer military service and were protected from being drafted by the Ministry of Labour. Satellite firms received no government compensation but were expected to be compensated by the nucleus firms. The Board of Trade approved three types of concentration: in some cases it was thought that a group of companies might agree that one of their number would be the nucleus company and manufacture products at cost for the satellite companies; satellite firms who had already been allotted raw materials could sell them on to nucleus companies; or satellite firms could be compensated by a levy on nucleus firms. In most cases, the first procedure was chosen.[507]

Being a satellite firm cannot have been a very appealing prospect, especially since it was for individual firms to negotiate the terms. As *The Times* commented,

> Although the Government has promised that they will do all they can to help firms deemed redundant to regain their former position after the war, members of all parties are wondering whether some of the things being done under necessity during the war can be undone.[508]

In the wool processing industries, it was not necessarily whole companies which were designated nucleus or satellite, but individual plants or departments. The Woolcombing Employers' Federation announced on 12 August that it intended to draw up a concentration plan for its sector. The Board of Robt. Jowitt & Sons resolved to enter into 'acceptable schemes of concentration for all three Branches'[509] and in October, George Harland signed: (1) a form of undertaking to the Woolcombing Employers' Federation agreeing to become a party to the scheme of concentration; (2) a form of application for membership of the Central Concentration Association; and (3) a supplementary letter addressed to the Central Concentration Association in respect of the company's business as topmaking woolcombers.[510] At the November meeting, C. S. Tullie told the Board that,

> applications had been sent in for membership of the woolcombing and wool carbonisers' schemes, these being subsidiary to the Central Concentration Association... It was anticipated that the Fellmongery would come under a national trade scheme, within the next few months.[511]

An application for nucleus status in respect of the Combing Section was read to the Board. In the following month, Sam Harland told the

Board that a letter was expected from the Board of Trade regarding the company's application for nucleus status for the Carbonising Branch, and confirmation of nucleus status for the Combing Section was also anticipated. The concentration scheme for fellmongery was still being worked out.[512]

Although it is not stated in the minutes, it is clear that the company achieved nucleus status for all its processing branches, since it continued processing. Nonetheless, it (and presumably all other wool processors) suffered serious rationing in all but the Carbonising Branch, and the Managing Director stated that combing rations were so low that there was no margin to be made. It was proposed to transfer staff from other branches to Carbonising in order to avoid the loss of skilled labour which would otherwise result.[513] In June 1943, the half-yearly accounts showed branches working at the following percentages of basic capacity: Combing 40%; Carbonising 50%; Scouring 25%; Sorting 32.5%; with Fellmongery temporarily ceased.[514] By October, Fellmongery had restarted and Carbonising had been augmented with commission work.

FELLMONGERING AT ADELAIDE

The fellmongering business at the Highbury Works, Meanwood, had suffered in the years leading up to the war and the mill had been put on short time, and even closed for periods. In contrast, fellmongering at Hindmarsh, Adelaide, had been a considerable financial success, partly due to lower wages, and it was decided to extend the plant in 1932 to increase drying room and warehousing space.[515]

The fortunes of the British fellmongering plant improved during the early part of the war. In January 1940, the mill was said to have three months' work, with more anticipated from government commissions.[516] The Wool Control took over the distribution of sheepskins in the UK, placing the company at the mercy of rationing, to the extent that the Meanwood fellmongering plant came to a temporary stop in mid-1943 due to a supply shortage.[517]

There were no such constraints on the Australian fellmongering operation, offering Jowitts an opportunity to build up its activity at Hindmarsh, Adelaide. Sam Harland advised the board that, given the relative appraised value of scoured and unscoured skinwool, it would make sense to install a four- or five-bowl scouring set. The Board approved the sug-

160

Fig. 41 William Thomas Benson Jowitt, whose Chairmanship
was tragically brief.

gestion,[518] but the plan had to be abandoned when it was pointed out that they had a non-compete agreement with J. W. McGregor & Sons Ltd, who acted as their agents.[519] In August 1940, the company spent £A438 on a new drying shed.[520]

In January 1943, Sam Harland told the Board that variable temperatures in Australia caused deterioration of sheepskins, and it was agreed that a refrigeration plant should be installed at Hindmarsh to prevent this. It was anticipated that it would cost about £2,500.[521] Unfortunately, on the morning of 11 March 1943, the newly-improved plant caught fire. When firemen arrived at about 4.30 a.m., the buildings, covering nearly two acres, were alight. Fanned by strong winds, flames leapt 100 feet in the air. Adjoining factories with stocks valued at £1 million were threatened as fire broke through the western wall of the fellmongery. Forty firemen could do little but contain the fire. A 100,000 gallon water tank in the middle of the plant could not be accessed due to the heat. A newspaper reported,

> Not since the disastrous fire at the Dunlop Perdriau building in October, 1940, has there been such a fire in Adelaide.
>
> Today all that is left of the four-acre buildings is a £1000 boiler which was unharmed, and several small drying sheds. Blackened brick walls surround a pile of twisted iron and charred woodwork.[522]

Despite the level of devastation, the Board was told that it was reparable locally, the initial estimate being that it would cost £A32,000.[523] The fire insurance promptly paid £A22,878, but the lowest quotation for rebuilding was £A29,700, plus another £10,000 for a new lorry, refrigeration plant and complete modern pickling plant. With a new plant offering an estimated 75% increase in capacity, Jowitts clearly felt that rebuilding would make commercial sense. Including a sprinkler system, the final cost was about £A50,000. The plant finally reopened in September 1944, and insurance amounting to £A26,844 for consequential loss was paid in January 1947.[524]

BOARD ROOM CHANGES

Mr A. T. Keeling, whose plans had formed the basis for turning Robt. Jowitt & Sons into a limited company, died on 30 April 1940 at the age of seventy-two. The Chairman, W. T. B. Jowitt, expressed the Board's appreciation for the wonderful work which Mr. Keeling had done for the company. Sam Harland added a tribute to Keeling's 'broad professional and commercial outlook and... sterling personal qualities'.[525] With his death, there seemed little point in retaining expensive London solicitors, especially with the difficulties of wartime travel, and the Bradford firm of Wade, Tetley & Hill took over from Trower, Still, Parkin & Keeling in 1941, 'for any local business,'[526] which in practice seemed to mean almost everything.

William Thomas Benson Jowitt's period as Chairman was brief, stretching from the end of 1939 until his death from a kidney infection on 19 February 1941 at the age of thirty-nine. Mr G. C. Veale expressed the very deep regret felt by himself and the Board at the death of the Chairman, 'whose presence was always an inspiration, whose personality was so attractive and whose loss nothing could replace'.[527] The new Chairman was his twin brother, Robert Benson Jowitt II. However, he was absent due to illness from early 1944. J. H. Jowitt's son, Frederick Robert Benson Jowitt, was elected to the board on 5 March 1941.[528] He did not have the £10,000 shareholding required by article 81 of the Articles of Association for him to be appointed a Director of the company, but this was waived by the Permanent Directors (Sam Harland, Robert Jowitt II and E. M. Jowitt), using the power vested in them under the same article, on 6 March 1941.[529]

E. M. Jowitt and Robert Jowitt II were seldom present at board meetings during the war.

THE ECONOMIC CONSEQUENCES OF THE WAR

As early as 1943, the industry was planning for post-war reconstruction, with the various trade bodies building up reserves for this purpose and Jowitts elected to contribute an initial subscription of £52 10s 0d through the British Wool Federation.[530] It later transpired that they had been paying levies to a number of different bodies, including the Woolcombing Employers' Federation and the Wool Carbonisers' Federation. Confusion about what levies were due was soon sorted out and £30 15s 5d had been paid to the Woolcombing Employers' Federation by the July Board Meeting. At the same time, it was noted that a small further sum would also become due through the Wool Carbonisers' Federation.[531] No-one could predict the future shape of the industry, but the company's Directors were not slow to realise that there would be a move to processing at source in the colonies,[532] something which would have a major impact in the post-war years. The industry also prepared for the end of the war by compiling detailed statistics on what machinery repairs and replacements would be needed after the war.[533] Since even urgent repairs were often prevented by shortages of materials and labour this would be an essential aid in getting the industry back to work. Separately, the Board of Robt. Jowitt & Sons wisely took the precaution of setting money aside in a Deferred Repairs Account to be used when conditions permitted.[534]

Before the end of the war, the company was also considering the labour position in its mills and the 'nucleus of skilled personnel which would be necessary'.[535] It also planned to contact its former overseas agents,

> in order that old contacts might be renewed where considered advisable, or new contacts made, as soon as export trade could be revived.[536]

A remarkable aspect of the company's wartime performance is that despite wool rationing, disruptions and staff shortages, work continued almost in a normal way, wages were increased and dividends were paid. Hardly a Board Meeting took place without financial performance being described as 'very satisfactory'. This is not to say that the company produced large profits during the Second World War, let alone the 'excess' profits of the Great War which had so upset Frederick McC. Jowitt. After profits of about £34,000 in 1939 and £42,000 in 1940, they were below £20,000 for the duration of the war (see Appendix X).

The Early Post-War Period, 1945–59

After consultation with the Central Price Regulation Committee I am
making an Order fixing maximum prices for women's ready-made non-
utility outwear. For each main type of ready-made garment there will be
two maximum prices, the higher of which will apply to any garment and
the lower to garments made from cloth of limited prices. Thus, the maximum retail price of a wool overcoat or costume (including purchase tax)
will be £20; if made from cloth weighing not less than 10 oz. a yard and
costing 12s. a yard or less (manufacturer's price), the price may not exceed
£14.

Stafford Cripps, 19 November 1945[537]

Since the Germans were more concerned with disrupting the production of armaments than garments, neither Bradford nor Leeds had been
a target for enemy bombing.[538] As a result, the local infrastructure suffered only from a lack of routine maintenance. However, if the managers
of Robt. Jowitt & Sons had imagined that things would rapidly return to
normal, they were to be disappointed. A country so long focused on the
prosecution of war could not easily revert to peacetime conditions. Labour continued to be in short supply well into 1946 as military and Wool
Control personnel were slowly demobilised and it was only on 15 May of
1946 that the company was freed from the restrictions of the Essential
Work Order.[539] Such were the labour shortages that an advertisement for
labour in the combing mill in 1946 did not elicit a single response.[540]

The vast government apparatus which had regulated almost every aspect of commercial life during the war was slow to relinquish control;
indeed, the new Labour government under Clement Attlee consolidated
state control by nationalising the coal industry in 1946, electricity supply
in 1947, the railways in 1948 and gas supplies in 1949. In addition, rationing of petrol and many foods did not end until the fifties. Nonetheless,
1946 saw a remarkable bounce-back in profits for Robt. Jowitt & Sons,
from £19,500 in 1945 to £100,400. By 1948, profits had reached £460,200
(Appendix X) on the back of rising wool prices. However, it was feared
that they were unsustainable and by March 1949 Charles Scott Tullie was
warning the Board of the high surplus stock levels (9,327 packs with an
estimated value of £725, 861) and 'dangerously high level of prices'. There
was always a difficult line to tread with regard to stock; the company

165

needed enough to ensure that its plant did not lie idle while avoiding large surplus stocks which might be written down. The Directors agreed that if high prices and high taxation continued the policy should be to have a restricted surplus of stock by the end of June,

> the objective being to clothe the Company's own machinery, leaving outside processing to look after itself; not to make profit the main objective, but to guard against the possibility of loss, in so far as it is possible to translate such policy into practical terms.[541]

By the following month surplus stock had grown to 22,653 packs with an estimated value of £1,598,683, although J. B. Harland said that it would still be necessary to buy 4,000 bales to provide continuity in the combing mill from June until the arrival of the new season's wool. He added that spinners were holding off purchasing in expectation of a fall in prices.[542] The fear of an imminent collapse in prices proved unfounded and by July the surplus stock had been reduced to 8,926 packs estimated at £626,384, with prices paid by spinners firming considerably (although the stock was valued at a similar price to before).[543] The year ended with a trading profit of £160,600, much lower than that for 1948 but still very acceptable. Despite electricity cuts to the combing mill, a shortage of commission carbonising and a truncated financial period (21 November to 31 October 1950), the company put in another excellent financial performance in 1950 (see Appendix X).

Although the 1950s were generally a time of economic expansion and optimism in Britain, the same problems which had bedevilled the wool trade for centuries still caused great unpredictability in the industry. Demand for services such as commission carbonising could slow, as it did in October 1950,[544] wool prices could rise steeply, as they did at about the same time,[545] only to decline by 66% the following year. Fellmongery was often slow, either due to shortage of skins or to receiving very mixed batches of skins.[546] The fluctuations were so severe that at the end of the first quarter of 1951/2 the home departments were reporting a loss of £34,423,[547] which was reduced to £1,396 a mere two months later.[548] Three months later that had increased to £6,961, but was more than offset by increased commission charges in Australia and South Africa of £34,927.[549] Although the year finished with a very satisfactory profit of

Fig. 42 Frederick Robert Benson Jowitt (1892–1965), the long-serving manager of Hollings Mill. He took over the chairmanship when Robert Benson Jowitt II resigned in 1948 due to ill health.

nearly £161,000, such swings in prices and availability made planning for the future very difficult.

Even when trade was going well, the high cost of financing stock meant that temporary increases in the bank overdraft were often needed. Barclay's granted an increase to £850,000 in February or March 1953[550] only to have to raise it to £950,000 in May. This brought the company's borrowings above the £1,500,000 authorised by its Articles of Association. Although this situation was not expected to last for long, the company was obliged to consult shareholders. Richard McC. B. Jowitt expressed his concern about the high borrowing in view of the company's avowed policy of avoiding undue speculation. The Managing Director, G. E. Harland, explained that of the total stock value of £2,562,650 the only surplus stocks not contracted for were worth £583,279.[551] By August stock in hand was worth £1,844,019 and stock not contracted for was down to a value of £234,146.[552] By October, uncontracted stock was back up to £495,429.[553] Despite the distractions caused by high borrowing, the year's performance was justly described as 'very good indeed,' with an aggregated profit before bonuses, commissions taxes and dividend of about £482,000.[554] The trading profit after these items was taken into account was £285,100 (Appendix X).

By the end of 1954, doubts were being expressed about the future viability of the Meanwood fellmongery, but it was felt that concentrating on quality rather than quantity might help. Stock write-downs of £172,000 reduced the 1954 profit before bonuses, commissions and taxes to £121,900.[555] The trading profit after these items was taken into account was £97,100 (Appendix X).

The stock situation in April 1955 continued to look worrying with the company holding £1,827,091-worth, with a surplus valued at £765,272. At the same time, the British Wool Federation was planning to release surplus stock onto the market. Combing was on a par with 1954 but fellmongery was down due to skins being more expensive than greasy wool. Carbonising output was much reduced and its prospects were described as poor.[556] Stock levels in May 1955 were slightly up but it was argued that demand for quicker deliveries obliged the company to hold large stocks. It was thought that a 'buying wave' would result if confidence returned after the general election.[557] The Conservatives were returned with an increased majority. However, the country was hit by a prolonged dock strike and the looked-for rise in values did not occur, necessitating a write-down of £169,000. The financial performance of the company was very poor in the year to 31 October 1955 with combing hit by sickness[558] and the Meanwood fellmongery suffering from unsuitable skin types.[559]

Continued sickness in the combing plant up to April 1956 hit output, while high demand from Japan made it difficult to buy the better wools at reasonable prices.[560] The Carbonising Branch at the Highbury Works had been losing money for a considerable time and the Managing Director, George Harland, proposed scrapping it and moving the scouring set there so that fellmongery and scouring would be on the same floor as one unit. He had commissioned a report from an architect, Mr R. B. Armistead, into the condition of the concrete floor, to establish whether it could support the scouring equipment. It was said to be 'rather grim'. On the other hand, if they continued carbonising a new acid bowl would be required.[561] It was decided in June to dismantle the scouring plant and proceed with the plan. This entailed remedial work on the floor and the bricking-up of supporting arches.[562] Despite a number of difficulties, the year to 31 October 1956 showed a far better profit than the previous year, at £219,000 before bonuses, commissions, tax and dividends (£119,000 after deductions).[563] This improvement was due to a stock write-down of 'only' £26,000.[564]

The new financial year started well, with revenue earnings in the two months to 31 December 1956 amounting to £25,792, an appreciable increase on the same period in the previous year. However, there were some concerns; although the Combing Department was achieving increased production, the 'tear' (the ratio of tops to noils) was unsatisfactory.[565] The fellmongery had slowed production due to high stocks of wool and high prices for skins.[566] In February 1957, it was reported that combing output was improved but fellmongering and scouring were down on the same period of 1956. Stock levels were still high with a surplus of 3,791 packs valued at £413,191 out of a total stock of 17,626 packs valued at £1,938,198. The overdraft at the English, Scottish & Australian Bank in Sydney had been increased to accommodate increased purchasing for the Japanese account.[567] By June, total stock, including 109,000 lb of sheepskins, had grown to 20,485 packs valued at £2,481,149. Of this, surplus stock accounted for an alarming 9,576 packs valued at £1,153,193.[568] With substantial buying for the Japanese market and high prices, it was decided in July that article 108 of the Articles of Association would need to be amended to permit increased borrowing.[569] At no point during the year did the Board show undue concern about the high stock levels, but the final figures for the year to the end of October 1957 showed a stock write-off of £284,000, leading to a massive loss of £201,298 before dividends, taxation, etc. £170,000 was transferred from the Reserve for Contingencies to the Profit & Loss account.[570] Rather unconvincingly the Directors consoled themselves with the idea that the whole industry was under strain, rather than that the losses had been caused by their failure to control surplus stock levels.[571] One possible solution to stock losses was the futures market, although George Blackwell's example naturally made them cautious. J. A. Jowitt suggested in June 1958 that the company should join the London Wool Terminal Market. This was rejected by the Board but it was agreed that the Managing Director and Secretary would investigate the matter and run a pilot scheme on paper.[572]

In July 1958, it was reported that revenue earnings were well down on the previous year due to large Dominion losses. On the positive side, combing production was maintained and fellmongering activity had increased, thanks to the absence of Mazamet buyers in the Australian market, which had sheepskins of the right quality available at reasonable prices.[573] At the beginning of October, as the company's financial year drew to a close, both combing and fellmongering were performing better than in the same period of 1957 but revenue figures were far worse. Surplus stocks were high at 7,359 packs valued at £543,341 out of a total stock of 16,901 valued at £1,337,258. It was agreed that the company would buy cautiously, following the market down.[574] Although better than 1957, the 1958 stock depreciation was still a substantial £135,000, dragging the company into another loss, and £130,000 was transferred from the Reserve for Contingencies to the Profit & Loss account.[575]

REBUILDING AND RENOVATION

Both manpower and raw materials had been severely restricted during the war with the result that plant and the buildings were in a poor state of repair. From 1945 until almost the very end of its existence, the company had to spend heavily just to maintain plant in a safe and productive state. On top of basic repairs and replacements, further expenditure was required to modernise and remain competitive against foreign and domestic competition. Thanks to the Board's far-sightedness, it was possible to pay for much of the fifties expenditure out of the Deferred Repairs

Account built up during the war. Expenditure on plant and equipment was also helped by the Income Tax Act of 1945 which replaced the old wear-and-tear allowances with an initial capital allowance of 20%, an annual write-down allowance of 25% and a balancing adjustment (either allowance or charge) when machinery was sold.

By 1944, it had been possible for some of the most urgent repairs such as the roofs to the Highbury Works and the Combing Shed to be carried out, although the latter had to be carried out in sections. Government control of both labour and materials through a licensing scheme delayed the repair of the rest of the Hollings Mill's roof until after the war.

New carding machines had been ordered in 1944, under the so-called 'A' scheme which provided for replacement of machinery 'as early as practical at the price ruling,' having priority over 'B' scheme supplies.[576]

The company's proposed refurbishment of the combing plant extended far beyond like-for-like replacement. Shareholders at the AGM in January 1947 were told that contracts totalling some £60,000 had been placed, with £25,000 allocated to replacing the old 54" wooden cylinder cards with 16 60" metal cylinder Platt Bros cards. It was also proposed that the whole combing shed should be converted from steam power to electricity at an estimated cost of £10,000, a step which was almost inevitable given problems with the water supply (see p. 172).

The first four of the new cards were scheduled for delivery in 1947 but, with the inevitable post-war supply problems, the first three were delivered in September 1948, with the motors following shortly after. The last of the cards was installed in December 1949.[577] They had been delayed not only by shortages of materials, labour and fuel but also by changes of specification caused by new Factory Acts and the electrification of the combing shed and other plant. By March 1950, the benefits of the new cards were evident.[578] The first five of the old cards were sold for £200 each but the rest could only be sold as scrap for £17 each.[579] Further improvements were made to the mill, including the installation of a new backwasher in 1949.[580]

New lighting was installed in both Hollings Mill and the Highbury Works in 1947 and a staff canteen was installed in Hollings Mill that year, charging one shilling per meal.[581] This was followed by renovation and redecoration of the Highbury Works canteen in 1949, which was by then twenty-five years old and completely out of date.[582] The canteens caused an unexpected dilemma when the Industrial Staffs Canteen Wages Board proposed that canteen employees should receive twelve days paid holiday – more than received by other workers at the Highbury Works or Hollings Mill.[583]

Repairs and the installation of new equipment were effected as soon as supplies and labour would allow. In the twelve months from 7 June 1950, Hollings Mill had redundant shafting taken out, the office moved, a tool grinder and a new capacitor installed, an old engine removed and a moisture-testing electric oven installed, enclosure walls built for the coal stacks and new stabling found for the horse. During the same period, the Highbury Works installed a new conveyor and a new drying machine, repaired the access road at a cost of £896 10s 0d, looked to replace the crushers which dated back to about 1921 and the willeying machine, installed a coal elevator, and proposed to spend a massive £4,000 lining the coal pits to prevent leakage.

Over the next few years at Hollings Mill, new strong boxes were obtained,[584] false twist motions were fitted to the second finisher gill boxes and a new scouring bowl and dryer were ordered from Petrie & McNaught for £15,061.[585]

The Adelaide fellmongery installed new heating equipment and the Sunbridge Road head office installed a fire alarm, resurfaced the garage yard and converted part of the first-floor warehouse into a typing room, kitchen and rest-room.[586] All this, and a continual round of painting worthy of the Forth Bridge, was just enough to keep the company running. In future years, increasing capital expenditure was required to modernise and innovate.

Towards the end of 1959, the Directors were confident enough in the future of British combing to contemplate a new twenty-comb plant, either building a new mill in a distressed area with the help of government grants or fitting out a vacant mill.[587] This plan was later dropped in favour of re-arrangement and expansion of the existing premises as the company's circumstances excluded it from funding from the government's 'Industry on the Move' scheme.[588]

Electrification

Although the company had shown itself forward-thinking and very willing to invest in new equipment, it had not seen the need to move from steam-powered to electrically-powered machinery. Indeed, at a time when power cuts were common, reliance on an external electricity supply was not particularly appealing – at least with a steam powered system, the mill manager could see that coal stocks were running low and plan the work accordingly. The company might have continued to rely on steam power for some more years but Nature and Bradford Corporation forced the company's hand.

In the Middle Ages, a mill race, known as the Goit, had been constructed to power manorial corn mills with water from the Bradford Beck. It was this Goit which still provided water free of charge for Jowitts' boilers in Sunbridge Road at the end of the Second World War. Unfortunately, heavy rains would occasionally cause flooding in the Goit upstream from Hollings Mill, and Bradford Corporation, faced with mounting claims for compensation, decided to prevent further flooding by blocking the entrance to the Goit. When this information was communicated to the Board in April 1945, it immediately placed the matter in the hands of Wade & Co., solicitors, the company noting that the Goit had apparently been 'available for such purposes since 1760'.[589] Another affected mill, Richardson's in Tetley Street, also made representations to the Corporation.[590]

One option considered by Jowitts was the installation of a cooling tower which would allow an almost closed system, requiring far smaller water supplies. The company was also open to the idea of electrification, but it turned out that a suitable mains supply was not available in 1945. Jowitts offered to pay £75 per annum for the upkeep of the Goit until a suitable supply became available.[591] While the Corporation investigated the possibility of supplying electricity, Robt. Jowitt & Sons consulted Brooks Electrical Motors Ltd and the British Electric Company about possible electrification schemes. It was acknowledged by the Board that 'electrification would mean improved conditions with regard to air space and cleanliness.' It was also noted that spacing between machines would have to be increased to comply with the new Factory Acts.[592] While Sam Harland was able to secure an electricity supply contract in mid-1946,[593] electrification plans for the mill proceeded quite slowly and it was necessary for the company to persuade the Corporation to keep the Goit open for another year.[594] By late 1947, the Corporation was determined to close the Goit but the electrical motors for the combing shed were not expected until 1948.[595] Jowitts was advised by Wade & Co. that no prescriptive rights could be traced which would force the Corporation to continue the supply. However, Bradford Corporation was in no position to supply the whole of the company's water supply. As a stop-gap measure, the company investigated the use of a well in the mill yard, and initial investigations suggested that this would provide a suitable substitute at up to 6,000 gallons per hour. For this to be viable, the company would also need permission to continue discharging into the Beck via the Goit (which would not be blocked downstream from Jowitts).[596] Further tests found that the mill could work on 4,500 gallons per hour and that water could be pumped from the well at 6,500 gallons per hour. Bradford corporation gave permission for the company to discharge water into the Goit downstream from the mill[597] and in January 1948 sealed it off upstream.[598] Unfortunately, the well water proved to be very hard,[599] causing scaling of the boiler. This was reduced by mixing it with 400 gallons per hour of water from the main Bradford supply. The electrical switchboard was functional by July 1948 and was found capable of handling lighting as well as power to the machinery.[600] However, it was agreed with Bradford Corporation that full electrification would not be completed until Spring 1949.[601] Unfortunately, Bradford's electricity supply was still inadequate and it was agreed that, by deferring full electrification to Spring 1950, Jowitts would avoid having to work staggered hours (which would otherwise be imposed by the electricity authority to achieve load balancing).[602] In the meantime, the supply to the electrified parts of the combing plant proved unreliable:

> Some stoppage had occurred as a result of electricity cuts, and the danger to life and limb had been pointed out to the Authorities, who were not prepared to give any warning, nor, up to the present, to allow a switch over to the remaining D.C. current available, for lighting purposes.[603]

Power cuts were still causing problems at the end of 1950 and Hollings Mill came under the staggered hours scheme in 1952.[604] Nonetheless, electrification had been unavoidable. Coal supplies were erratic and it was found that the mill's well ran short of water in dry conditions.[605]

Micro-Management?

The directors' careful oversight of capital expenditure was commendable, but there was, perhaps, a tendency to micro-manage, with the

Board discussing quite small items of expenditure such as an electric typewriter[606] and a Sumlock adding machine.[607] Scrutiny of Australian expenditure was every bit as close, with car purchases and even much smaller items needing approval from the Board.[608] Cars for UK managers were a constant source of discussion at Board meetings but they were in very short supply during the fifties as the government tried to rebuild its foreign currency reserves. As a result, Jowitts found it worthwhile buying British cars in South Africa and shipping them back. Even new vegetable sinks for the canteen required approval at board level.[609] This tight rein on capital expenditure was in marked contrast to the company's occasional and costly failures to control reckless behaviour by agents and wool-buyers at home and abroad (see, for instance, pp. 188, 199, 217).

THE NEW LONDON OFFICE

A new London office was leased in the Wool Exchange, next to Spitalfields, from March 1946 at an annual cost of £164 12s. The company also put down a deposit on 12 Maple Avenue, Leigh-on-Sea, as a residence for the returning London Manager, George Eke.[610] He would live there rent-free for the duration of his employment as Manager. His removal expenses were also paid by the company. The house was sold on behalf of the company by Peter J. M. Bell for £3,500 in 1965.

CHANGES TO SHARE CAPITAL & ARTICLES OF ASSOCIATION

With a very substantial overdraft,[611] the company was interested in raising fresh capital. However, the existing Articles of Association and the structure of the company's share capital were unfair to shareholders who were not close family members and also made the company an unattractive investment to trusts and insurance companies. In order to facilitate the sale of shares to these organisations, a capital restructuring, with the creation of 6% cumulative preference shares, and changes to the Articles of Association were proposed. A public issuing of shares was contemplated because it was felt that the prospect of flotation was necessary to attract institutional investors,[612] although some Directors, notably Robert Benson Jowitt II, were implacably opposed to any loss of family control.

An Extraordinary General Meeting was called for Wednesday 24 March 1950 at which proposals for a restructuring of the share capital were placed before shareholders. It was proposed that the existing capital, consisting of 250,000 fully paid-up £1 ordinary shares, would be increased and re-organised into 250,000 6% cumulative preference shares and 250,000 £1 ordinary shares. This re-organisation was to be achieved by: (1) converting each existing £1 ordinary share into five shares of 4/- each, two of which would be ordinary shares, and three preference shares; (2) by consolidating the resulting 750,000 4 shilling preference shares into 150,000 £1 preference shares and the 500,000 4 shilling ordinary shares into 100,000 £1 preference shares; (3) by creating a further 100,000 £1 preference shares and 150,000 £1 ordinary shares.[613]

The company then turned its attention to the Articles of Association which had been drafted in 1919 and were out of step with subsequent legislation. A committee was set up to investigate.[614] The original company structure which had served Robt. Jowitt & Sons well since incorporation was showing its age by 1950. Restrictive clauses in the original Articles of Association gave Permanent Directors the right to veto the transfer of shares and the right to appoint Ordinary Directors. This allowed successive generations of the Jowitt and (to a lesser extent) Harland families to retain control of the company, but meant that someone who inherited shares in the company would have difficulty selling them to anyone but a family member, yielding what might be an unreasonably low price. Indeed, no-one who inherited shares could even have them registered in their name without permission from the Permanent Directors.

The changes to the Articles of Association, approved at an Extraordinary General Meeting on 14 March 1951, removed the right of directors to restrict the transfer of ordinary shares, while retaining this right in the case of the recently-created 6% preference shares. The new Articles also removed the concept of Permanent Directors, who were the nominated successors of previous directors. A further clause specified that those who inherited or otherwise became entitled to preference shares would be entitled to the same benefits as other holders of preference shares, except the ability to exercise 'any right conferred by membership in relation to Meetings of the Company until he shall have been registered as a member in respect of such share'. Holders of the preference shares would also be paid first in the case of liquidation or any other return of assets to shareholders.[615] To mitigate the loss of guaranteed family succession, the Board entered into a gentleman's agreement (i.e. binding in honour but not in law) allowing Robert Lionel Palgrave Jowitt to be appointed an Ordinary Director immediately upon his retirement as a Permanent Director. Similarly, it was agreed that Maurice's son John Alan Jowitt should be

> nominated as an Ordinary Director to fill the present or any subsequent casual vacancy as and when the Board deemed it desirable to elect a member for this purpose.[616]

The changes to the Articles of Association were approved at the EGM, although some further changes were made to them at an EGM on 14 March 1951. A number of institutional investors were subsequently accepted as shareholders in the company.[617]

The Articles were altered at an EGM once more on 23 October 1957 when, by special resolution, Article 108 was replaced, allowing the directors to borrow up to £3 million rather than £1.5 million.

FINANCIAL YEAR END

Before Robt. Jowitt & Sons was incorporated in 1919, the partnership had finished its financial year on *approximately* 20 November (there seems to have been some variation). With incorporation, that date was adopted as the official year end, although overseas branches worked to different dates. In January 1950, the Board was informed that the new Companies Act obliged them to have accounts of the Head Office and all branches made up to the same date. It was decided that, subject to any adverse tax implications in Australia, all parts of the business would adopt 31 October as their year end.

PENSIONS SCHEMES

Another overdue change was made in 1950. Since its incorporation in 1919, the company had shown a paternalistic care for its former employees by granting them pensions, but these were usually termed 'voluntary allowances,' and they were exactly that, voluntary payments made at the discretion of the directors. Even Joe Sheard who had worked for the company for sixty-two years did not receive a company pension as of right.

On 15 May 1950, the superannuation schemes for staff in the UK and South Africa came into effect with the Legal and General Assurance Society.[618] The cost for the UK and South African schemes had been calculated at £1,469 per annum for future pensions. Pensions for past services would be paid for by either a bulk payment of £34,745, spread over eleven years for tax relief, or eleven annual payments of £3,646.[619] At the same the Australian Mutual Provident Society was asked to come up with a scheme for Australian employees along exactly the same lines[620] and that scheme came into effect on 27 October 1950.[621] A superannuation scheme for directors and senior managers was arranged with Noble Lowndes Pension Services in August 1951.

Although the pension scheme could be seen as quite generous and forward-thinking, it only applied to salaried staff and so-called 'standing-wage men,' defined by the company as those who are,

> paid a full week's wage, whether actually working or not, e.g. during periods of sickness; and the scheme shall be deemed to include Foremen, Overlookers, Maintenance Staff, and Others who are paid on the standing wage basis, and who do not receive payment at overtime rates for any work done during non-running hours.[622]

The company had considered including the remaining employees, classed as 'operatives,' in the pension scheme, but eventually decided to exclude them.[623]

For many years after the introduction of the superannuation scheme, voluntary allowances continued to be paid to existing pensioners, and also for those former employees who didn't qualify for the scheme. The allowances were also used to augment the superannuation scheme, as in the case of Mr J. S. Yarborough who had been off sick since suffering a stroke in February 1952 and was awarded a voluntary allowance of £100 p.a. in addition to the pension he received on retirement on 15 May 1954.[624] Members of the scheme who retired soon after its inception had not had time to build up much if any entitlement and it was suggested that it might be best for some employees to defer retirement past the usual retirement age (65 for men and 60 for women) to give them an increased pension.[625] One such case was Mr W. Hall who had been with the company since 1906. He extended his employment until he was sixty-six and thereby completed one year under the superannuation scheme.

Fig. 43 James Charlton Reid.

He was granted a further £100 p.a. voluntary allowance.[626] Charles James Hunt, the engineer at Meanwood, who was due to retire at the end of 1952 with £109 18s 0d p.a. from the superannuation scheme, was granted an additional voluntary allowance of £100 p.a.[627]

BOARDROOM AND MANAGEMENT CHANGES

The early postwar period was marked by changes in management at the top of the company. On 6 September 1945, the Company Secretary, Charles Scott Tullie, and Harry Ingham, long-serving Manager of the Top Department were appointed Ordinary Directors. Ingham brought a wealth of practical experience to the Board and Tullie a fine financial mind.

At the end of 1945, W. R. Myers, Manager of the Port Elizabeth Office, sent a cable to Sam Harland informing him that he had been invited to become the General Manager of the Joint Organisation set up to sell off surplus Dominion wool accumulated during the war. G. E. Harland undertook to write to Myers asking him to arrange to train up James Charlton Reid as Manager.[628] Reid, who had joined the company in 1923, working under his father G. A. Reid in the East London Office, had only recently returned from active service in Libya, Egypt (including the Battle of El Alamein), Italy and France.[629] He was duly appointed Manager of the Port Elizabeth Office in mid-1946.[630]

The Jowitt family, in line with the founding principles of the company and its Articles of Association, continued to hold the permanent directorships and Harlands continued to hold other key positions on the

179

Board. Robert Benson Jowitt II, brother of the first Chairman, Frederick McCulloch Jowitt, died in Winchester on 5 December 1945. His involvement in the company had been limited during the Second World War, but Sam Harland, who had worked with him since 1903, praised his very active involvement in the pre-war period.[631] Under the terms of Robert Jowitt II's will, his trustees appointed his son, Robert Lionel Palgrave Jowitt, to the Board on 9 May 1946. At the same time, J. B. Harland, son of Sam Harland, and Robert Benson Jowitt II's brother Richard McCulloch Benson Jowitt were admitted to the board. R. L. P. Jowitt and J. B. Harland expressed 'their worthy desire to maintain the high standards of their predecessors,' but R. McCulloch Benson Jowitt was absent from his meeting, a fact which was regretted by Sam Harland who was in the Chair.[632]

The uneasy sharing of power between Jowitts and Harlands meant that members of the two families expected to fill important positions in the company and, if a suitable position did not exist, one would be created for him. A particularly unfortunate example of this was Jim Harland (Geoffrey Harland's brother) who, for some reason, did not take up the important role in the Australian business for which he had been carefuly groomed. As an alternative career path, he was allowed to set up a Firm Offer Department in 1963. It seems not to have occurred to anyone on the Board at the time that by offering fixed-price contracts to supply wool to other topmakers, the company would be undermining the profitability of its own topmaking activities.

Robert Benson Jowitt II, whose period of Chairmanship since 1941 had been marked by long absences due to ill-health, tendered his resignation in a letter of 2 October 1948. In what was possibly an overdue move, it was unanimously resolved to appoint the hard-working F. R. B. Jowitt as Chairman. In contrast to Robert Benson Jowitt II, he had attended nearly every Board meeting since 1941, as well as managing Hollings Mill.

The auditor, Albert Barker of Peat, Marwick, Mitchell & Co., retired on 30 September 1949. In a gesture which would nowadays cause a few eyebrows to be raised, the company gave him £500 as

a tangible expression of the very high regard and esteem in which the Board of Directors held him personally, and of the sincere appreciation of his valuable and conscientious work on the Company's behalf.[633]

R. L. P. Jowitt was due to lose his position as a result of the 1951 changes in the Articles of Association. In line with the gentleman's agreement of 7 June 1950, it was resolved by the Board on 7 January 1951 that he would be re-appointed immediately the new Articles came into effect. Frederick McC. Jowitt's widow, Helen Dorothea Jowitt, died on 12 October 1952. Although never a Director, she had been a major shareholder since the incorporation of the company in 1919.

Edward Maurice Jowitt had been unwell for some time and on 13 September 1953 he gave the requisite one month's notice of his intention to retire. Under the gentleman's agreement of 7 June 1950, he had nominated his son, John Alan ('Jack') Jowitt as his replacement on the Board.[634] Jack was duly welcomed as a Director in October 1953.[635] Taking after his great-grandfather John jr, he was a keen artist, studying under George Mitchell at Yeovil School of Art from 1935 to 1939. For many years, the company's calendars were adorned with his paintings.

Also in October 1953, Charles Scott Tullie, who had been Assistant Company Secretary from 1924 and then Company Secretary since George Blackwell's removal at the end of 1931, died during an operation at the age of fifty-four. R. B. Jowitt II, standing in for F. R. B. Jowitt who had been injured in an accident at Hollings Mill, asked the Board to stand in silence as a tribute to Tullie, 'a very dear friend and most worthy colleague'. Mr Tullie had £10,112 available under the superannuation scheme and the Board thought that part or all of it should be invested in an annuity for his widow. The Board agreed to ask Colonel Wade of Wade & Co. to offer advice on the matter to Muriel Tullie.[636] The company also made a gift of £2,000 to Mrs Tullie.[637] Douglas Penney Barker FCA, son of the former Peat Marwick auditor, Albert Barker, was appointed as Secretary starting 1 January 1954, at a salary of £2,250 p.a. plus bonus, and the long-serving Miss Bertha Tordoff FCIS was confirmed in her role as Assistant Secretary with an increase in salary of £250 p.a.[638]

Edwin ('Ted') Bell, long-serving salesman in the Topmaking Department, was appointed to the Board on 22 July 1953. It is likely that George Harland, the Managing Director, had heard that Lord Barnby had made a similar approach to him.[639]

Donald B. Sykes retired from his position as South African Director on 31 October 1953 after fifty years with the company and the Board resolved to augment his pension under the superannuation scheme to bring it up to £450 p.a. Mr Sykes recommended James Charlton Reid, Manager of the Port Elizabeth Office, as his successor,[640] which he accepted.[641] Sadly, Reid, an apparently fit man, hardly lived long enough to enjoy the new position. He died on 9 September 1954 of a brain haemorrhage after a round of golf with Donald Sykes's son Tom. Reid was only forty-seven.[642] Norman Skinner was made Manager of the Port Elizabeth Branch and D. B. Sykes returned as South African Director.

Both the former Managing Director, Sam Harland, and the former Chairman, Edward Maurice Jowitt, died on 22 September 1954. The Board stood in silence as a token of their respect for them, as well as for James Charlton Reid, whose sudden death was described as 'tragic'. Mr D. B. Sykes was asked to contact Mr Reid's widow to offer the company's assistance.[643] The Board later resolved to provide £150 per annum for the education of each of Reid's children up to the age of seventeen.[644] This was extended slightly in the case of the eldest son, David, who started as a trainee wool-buyer under D. B. Sykes's son Thomas Percy Sykes in 1959.[645]

Geoffrey Harland, the elder son of George E. Harland, the Managing Director, was appointed to the Board with effect from 17 March 1955.[646]

D. B. Sykes, the South African Director, offered to retire again to save the company money in September 1959. However, the Board, while considering that the economic position in the Cape did not really warrant a full-time Director asked Mr Sykes to retain his directorship in view of his very faithful service to the company.[647]

F. R. B. Jowitt, who had been suffering from cancer, resigned from the Chair at the end of 1959, although he remained on the Board. George E. Harland, the Managing Director, a heavy smoker, missed the September, October and November meetings due to ill health and died of lung cancer shortly after. His willingness to buy wool whatever the market conditions and irrespective of the company's stock position had caused a stock loss of £300, 000,[648] creating the limited company's largest ever losses in 1957. Cruelly, but perhaps accurately, it has been suggested that his demise saved Robt. Jowitt & Sons from disaster. His brother, James Burgess Harland, was appointed to replace him in December.[649]

The New Generation

Like many children at the time, Frederick Thomas Benson Jowitt ('Tommy') was evacuated to the safety of Canada during the war, where he had a rather unhappy time at Sedbergh School in Montebello, Quebec. His father W. T. B. Jowitt had died in 1941 and by the time Tommy returned to England in June 1945 his mother had remarried to a Lieutenant Commander in the Royal Navy. Tommy was sent to Wellesley House School briefly before attending Eton. His hopes of going to Trinity College, Cambridge, were dashed by his stepfather, a man of unswerving self-importance. Instead, Tommy was articled to the accountancy firm Peat, Marwick, Mitchell in 1952. As soon as he could, Tommy escaped from accountancy, joining the family wool business in 1957. He started in the Wool Sorting Department where he met another relative newcomer, Peter Bell (see below), before moving to the Woollens Department, supervised by R. B. Jowitt II. Tommy won a scholarship from Bradford Chamber of Commerce to study in Germany, where he spent six months becoming fluent in the language. It was to serve him well in his eventual role in the Export Department, where he had special responsibility for Austria, Southern Germany and Switzerland.

Entering Robt. Jowitt & Sons six months before Tommy was Peter J. M. Bell, son of Edwin Bell. Peter was well prepared for the technicalities of the wool trade, having received a Textile Diploma after three years' study at Bradford Technical College (now Bradford University), followed by two years' National Service. The Technical College could trace its origins to the foundation of the Bradford Mechanics' Institute in 1832. Like his father before him, Peter became a salesman in the Top Department and eventually took over J. B. Harland's responsibilities for wool stock control. In 1963, following F. R. B. Jowitt's retirement, Peter took over responsibility for Hollings Mill.

Peter and Tommy found themselves in a strange and rather bewildering environment for which neither Bradford Technical College nor Eton could have prepared them. Not only was there a deep-seated antipathy between the Harlands and the Jowitts, each department was a separate fiefdom, run not for the greater good of the company but for the aggrandisement of its manager. There was no proper attempt to identiify cost and profit centres with the result that totally uneconomic activities would be maintained at the expense of more profitable ones.

The entry of Peter and Tommy into the company was not welcomed by the Harlands, who probably feared some erosion of their power. While they could do nothing to block Tommy's progress, they made life difficult for the young Peter. Fortunately for him, R. B. Jowitt II's support for Peter's father secured his position in the company and Tommy's arrival provided a friend and ally. The two young men would go on to make far-reaching and much-needed changes to Robt. Jowitt & Sons Ltd.

THE STATE OF THE BRITISH WOOL INDUSTRY

Robt. Jowitt & Sons was not alone in its unsystematic approach to business in the 1950s. The whole Bradford wool industry, and indeed the British wool industry in general, was craft-based, unscientific – spinners, for instance, were still checking the tension of the threads by touch – and very often ridiculously over-specialised; there were companies which dealt only in the wool of one particular British breed of sheep or sold only daggings – the manure-rich wool from the hindparts of the sheep. The industry was also deeply secretive, tormented with paranoid fears that competitors would gain some advantage over them. In reality, despite secret codes and other subterfuges, most companies had a very good idea what their competitors were up to; it was well-known that most of the lorry drivers would happily discuss their routes over a pint or two.

OVERSEAS OFFICES

Throughout the post-war period, the Bradford Head Office seemed to display a rather ambivalent attitude to the Australian operations. To some extent this reflected rivalries between the Harland and Jowitt camps, but it also fluctuated with the unpredictable and ever-changing economic situation in Australia and Britain. In the forties and fifties, the Australian offices were hampered by labour shortages,[650] drought,[651] good weather (encouraging farmers to hold on to sheep for longer with resultant skin shortages at the Meanwood and Hindmarsh fellmongeries)[652] and exchange rate fluctuations.[653] On the other hand, a loss of £6,961 by the Home departments in mid-1952 was more than compensated for by Dominion profits of £34,927.[654]

At first, the company was intent on expansion, buying another fellmongery in Fremantle,[655] later described as 'a tin shed with concrete floor,' for £A2,200 in 1947 and opening a Melbourne Office. Land at Dry Creek near Adelaide was bought[656] and new offices were sought in Sydney.[657] However, by 1954 the Fremantle fellmongery had been declared unviable.[658] The Adelaide fellmongery, which Arthur Gibson had described as 'the best plant of its type in Australia' in 1946,[659] was closed indefinitely in 1956 due to skin shortages,[660] and again at the end of 1957,[661] followed by the decision to sell it off in January 1959.[662] An auction of the plant in 1959 failed,[663] but a sale for £A70,000 went through in January 1960.[664]

The Australian operation had always been about more than fellmongering. Indeed, its main purpose was wool-buying, both for the company's own use and on commission. Unfortunately, there were serious concerns about the quality of buying in Australia. In what might have been an isolated incident, Bradford clients made a claim against the company for underyielding wools in 1955.[665] However, this was just the first of many problems. In January 1956, feeling that Adelaide was paying too much for skins, an attempt was made to impose limits from Bradford, an experiment which was unsuccessful. As a result, it was decided that J. B. Harland and Jack Smith from the Meanwood fellmongery would investigate skin-buying in Adelaide.[666] J. B. Harland subsequently told the Board that there was a strained atmosphere among the Adelaide buying staff. He also noted that testing equipment, in good order, had fallen into disuse owing to 'labour difficulties'.[667]

A. A. Gibson retired as Australian Director on 26 January 1959 and the board were assured by George Harland that Gibson's successor, Simeon Myers, would 'keep a closer watch on the various Branches than had been the case in the past'.[668] It was a surprising remark, as there is no record that anyone had ever criticised Gibson's oversight since he became a Director in 1926. It probably reflects the partisanship which so often wreaked havoc in the management of the company. Gibson, who got on well with RBJ II and his Australian wife, Audrey (*née* Stanton), was perceived as being in the Jowitt camp, and it must be remembered that Gibson's family had worked for the Jowitts long before the Harlands joined the company. Myers, on the other hand, was related to Harry Ingham and was thought to be a Harland man. In May the Board asked Myers to investigate why so few skins were being bought.[669] The Board were still expressing concern over Australian buying at the end of 1959.[670] In fact, it was a problem which would keep recurring. As a money-saving measure, it was decided to merge the Perth and Fremantle offices (which were only a few miles apart) in May 1959. A plan to merge the Sydney and Melbourne offices was deferred.[671]

By contrast, while the South African operation experienced a number of problems, no-one seems to have found fault with the quality of its buying since the days of Victor Tate, although it has to be said that the quality of wool available in South Africa was markedly inferior to the best Australian wools. David Reid, son of the former South African Director, James Charlton Reid, describes the *sub rosa* operation of the South African branches:

> After School in 1959 I Joined Robert Jowitt & Sons as a Learner Wool Buyer. The wool trade in Port Elizabeth was very competitive but there was great camaraderie amongst the buyers. In those days wool sales were held every Friday with over 3,000 lots on offer. Each lot had to be valued for style, staple length, spinning count and clean yield. Every Thursday night Jowitts in Bradford plus US and Japanese clients would cable orders with clean yield prices. In real James Bond Style all this was done in code. On Friday at 6 a.m. I would collect cables from post box, decode and enter greasy prices into catalogue. For instance, if the price limit was 100 ¢/kg and you had put a clean yield of 60% on it, your max price at sale was 60 ¢/kg. Jowitt worked on a 5-letter word code system and numbers were RBJOWITANX (1 to 0). That is, 65% would be written with the letters IW.[672]

The code books sent from Bradford (which became obsolete some years later with the installation of telex machines) were the size of *Encyclopaedia Britannica* volumes and covered every eventuality. One single five-letter word carried the meaning 'cease trading and come home at once'.[673] This code word was sent to Jim Harland on the death of his father.

POLITICAL CONTRIBUTIONS

Although in the nineteenth century the Jowitts had been very active and vocal in political matters, usually being aligned with the Liberal Party, the limited company had always adopted a policy of refusing all political donations. An appeal for money from Bradford Conservative and Unionist Association in June 1939 was unanimously rebuffed because 'subscriptions could not be made to Political Organisations.'[674] A similar application from the City of Leeds Conservative Association in 1944 was dismissed, as it was agreed that 'political activities should be considered outside the scope of the subscription list.'[675]

The situation changed in 1947 when the Conservative and Unionist Party tried a different approach. Rather than an appeal from the local party, Jowitts received a request from Conservative Central Office for a donation to 'a special fund to fight for the preservation of private industry'.[676] In early post-war Britain under a Labour government it must have seemed to many that private industry was, indeed, under threat. Robt. Jowitt & Sons was at the mercy of Bradford Corporation for their (often inadequate and unreliable) electricity and water supplies, while rationed and erratic coal supplies were the responsibility of the recently-nationalised National Coal Board. Much of the heavy-handed, bureaucratic control of basic supplies and services which the company had accepted without complaint in wartime dragged on for years. It is hardly surprising that the company's attitude to political donations started to shift, especially as it was dressed up as something slightly different:

> In view of the Company's declared policy not to subscribe to political appeals a long discussion took place on the aims of the Special Fund referred to, and it was unanimously agreed that they covered the protection of the Company's interests as a private enterprise. Mr. R. B. Jowitt prposed, Mr. G. E. Harland seconded, and it was unanimously *Resolved* that a donation of £100 be given for this purpose.[677]

When the City of Leeds Conservative Association asked for a donation in November 1948 and 'funds for the protection of private enterprise' in 1949, the company did not hesitate to donate ten guineas.[678] Although the company never became a major political donor, the taboo had been broken, and donations to the Conservatives were made in 1954, 1958 and 1962 with no reference to a special fund.[679]

The Return of
Robert Benson Jowitt II

Directors of the 1,500 firms in the British wool textile industry are dismayed by the lithe young people who bound through newspaper and television advertisements proclaiming that their garments never shrink, wrinkle or crease because they are made of such-and-such a manmade fibre. A market survey has confirmed what the wool textile industry had feared – the smart folk who people the advertisement world in their sharply creased trousers and trim pleated skirts are associated by the great British buying public not with wool but with the synthetic fibres, such as Terylene.

The Times, 10 August 1961

Robert Benson Jowitt II returned in the the role of Chairman, replacing the ailing F. R. B. Jowitt at the beginning of 1960. It did not bode well that he was absent due to ill health for what should have been his first meeting on 20 January, when his place was taken by R. McC. B. Jowitt.[680]

Although the financial results for 1959 had been moderately good, the general trend was downward as the performance in 1960 seemed to confirm. In late April it was reported that revenue earnings to date showed a loss of £18,686, reduced by a small profit for the month of £2,794. In addition, the company faced lean margins and held £729,592-worth of stock which had not been contracted for.[681] On 18 May, the Directors discussed the movement towards shorter working hours which equated to higher labour costs. It was noted that there was a move towards running smaller units for longer periods using a three-shift system. Jack Jowitt suggested that the company might seek a regular market for some of its output by co-investing in spinning facilities, an idea which was to be taken further in 1961 with an Indian joint venture (see p. 204). Surplus stocks had grown in value to £916,938 and the profit for the month to date was a pitiful £316.[682] However, by June the Secretary was able to announce that the loss for the last seven months had been reduced to £27,000. The Board held a long discussion on the markets for wool and tops and the general consensus was that, with stocks in Bradford low, prices were sound. A price of 105d for 64's seemed a suitable guide for sales. The company held stock of 22,953 packs valued at £1,872,370, including surplus stock of 11,803 packs valued at £975,852.[683]

In July 1960, Harry Ingham expressed a wish to take things more easily. It was agreed that his position as topmaker would pass to Geoffrey Harland who had been working as his assistant since 1953. Harry Ingram agreed to remain in an advisory capacity and as a Director. The Board expressed its appreciation to Mr Ingham for his service of more than fifty years. In view of his liking for pheasant shooting, he was invited to select a gun up to the value of £150.[684] He was very much a Harland man and much given to Delphic pronouncements – but then he was from Pudsey, a town whose inhabitants are notorious for answering one question with another.

By July surplus stocks had grown to 14,578 packs valued at £1,151,066. Revenue losses to date were £30,633, compared with a profit of £45,342 at the same time in the previous year.[685]

The Company Secretary, Douglas Penney Barker FCA, was appointed to the Board in September 1960.[686] The revenue returns reported in this month showed an aggregated loss of about £55,000. Surplus stocks had declined to 9,759 valued at £730,811, reflecting a drop of about 5% in the average price per pack since July. The Managing Director, J. B. Harland, said that large stocks of wool, tops and yarn in all countries had depressed prices at the opening of the new season and they were not expected to rise consistently in the near future. Lower wool prices were not entirely bad as the company had an oversold position creating a shortage of 2,157 packs valued at £163,422.[687]

The draft figures for the year to 31 October 1960 were presented to the Board in December 1960, showing an aggregate loss of £98,000, later increased to £100,520, compared with a profit for the previous year of £118,000. £75,000 was transferred from the reserves to the Profit & Loss Account,[688] the same amount that had been placed in the reserves the previous year. The Board was unanimous in agreeing that

economies would have to be effected, and certain laxities which had developed during prosperous times would have to be tightened up. The Chairman pointed out that the best way to effect this was in the first instance by all the Members of the Board maintaining high qualities of leadership and exemplary standards of behaviour in all fields.[689]

The Directors also agreed to recommend to shareholders that no dividend be paid on the ordinary shares for the year ending 31 October 1960 but that, should the trading results justify it, an interim dividend should be paid in March in respect of the year ending 31 October 1961.[690] This decision was reversed in a rather macabre fashion when the Board learnt that the disgraced Company Secretary, George Blackwell, had died on 31 December 1960. The company, as the beneficiary of his life assurance policy worth £22,728 8s, as well as a more modest sum received from a policy on the life of Sam Harland, suddenly had the money to pay a dividend.[691]

By 1961, Robt. Jowitt & Sons was feeling the effects of competition from abroad in the wool market and from synthetic fibres at home. The prohibitive effect of US import duties on wool was also a concern to the company. J. A. Jowitt stressed the need to advertise all wool products, and it was suggested that he wrote to the International Wool Secretariat about this.[692] In April 1961, the company was facing a loss to date of £21,301, leading R. McC. B. Jowitt to point out 'the serious repercussion on the Company's standing, and on actual share values, of the continuous appearance of red figures on the revenue sheet.' The potential for stock losses was also high, but the company was by now using the futures market to limit such exposure. The Chairman, R. B. Jowitt II, reporting on a visit by Lord Barnby, said that at current selling prices there was no future for combing. Barnby was proposing some sort of combined selling organisation. It was decided to examine any concrete proposals which might be put forward but to proceed with caution.[693] In May the Board discovered that the proposal was in reality something along the lines of a horizontal merger with Lord Barnby's Aire Wool Company and other concerns, an idea unacceptable to the Directors, who wished to see Robt. Jowitt & Sons Ltd retain its separate identity. It was decided to buy one hundred shares in the Aire Wool Company through a nominee so that the company would receive notice of any proposed mergers. The May figures showed a small decrease in the loss to date,[694] but the loss had increased to £25,272 by July. The Top Department held surplus stock of 4,873 packs but 80% of this had been hedged on the term market. Perhaps accepting that combing in the UK was no longer viable, the company decided to invest in an Indian combing plant in order to generate turnover for the Australian branches (see p. 204).[695] The trading figure for 1961 showed a gross profit of £40,865.[696]

Chairmen, Managing Directors, Directors and Company Secretaries had come and gone since the incorporation of the company in 1919 but

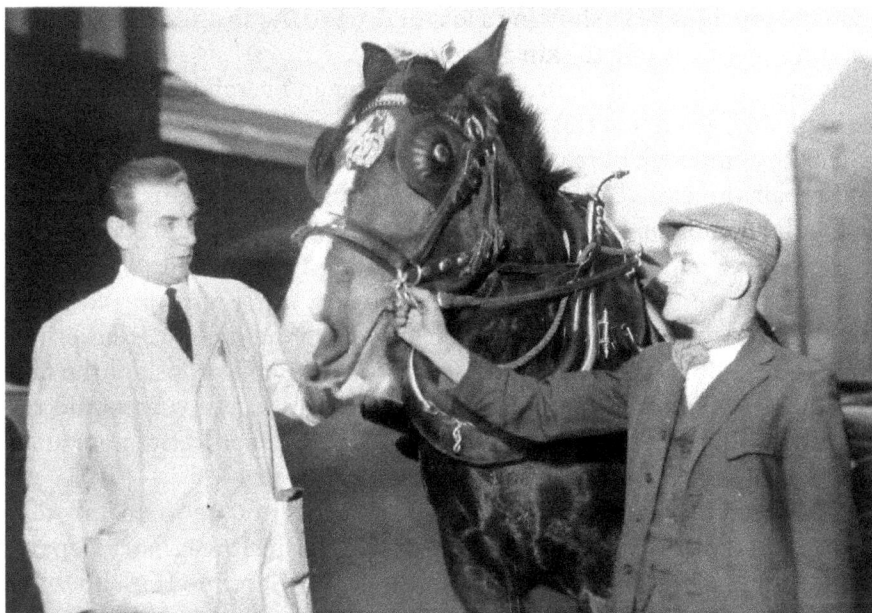

Fig. 44 'Boy', one of a long line of horses employed at Hollings Mill, November 1962.

one person had been present at board meetings throughout. In September 1961, Miss Bertha Tordoff, the indefatigable Assistant Secretary, retired after forty-two years.[697] Strangely, there is no record of what gift the company made her on her retirement but she had been a much-valued member of staff, and staunch supporter of the Jowitt directors as opposed to the Harlands, and there can be no doubt that the Board members expressed their appreciation generously. Also in September, it was agreed that Harry Ingram should have his contract as an executive renewed for another year, until 31 October 1962. He would remain a Director after this date.

1962 was a far more difficult year than 1961. The trading figures looked satisfactory in January but the company was owed £3,563 by Illingworth (Wools) Ltd, whose Managing Director was a close friend of Geoffrey Harland's, with little prospect of payment.[698] By March, the trading account was recording a loss due to low buying in the overseas branches. Revenue for the six months to 30 April 1962 were poor, though better than the previous year, showing a loss of £13,240. As so often, Meanwood was on short-time due to skin shortages.[699]

By July, the revenue figures were showing an overall loss of £25,700. Remarkably, given that this was the age of the 'white heat of technology,' when the company's horse was retired, it was decided to buy another one and repair the cart.[700] With a continuing lack of sheepskins from Australia, the Meanwood fellmongery was still working at reduced capacity in September and the question of substituting English skins was raised. Revenue figures for the first ten months were worse than in 1961, and the Chairman remarked that nearly everyone in the trade was suffering. Nonetheless, it is typical that the Board spared time to consider the fate of Hollings Mill's old horse. It was agreed that the company would retain ownership, but the horse would go into honourable retirement at an equine rest home, which would be given a gift of £100.[701]

At the end of 1962, the company showed a loss of £9,000, mainly due to losses in the overseas branches. It was also felt that the Export Department should be more flexible and Ted Bell and Geoffrey Harland were asked to investigate its possible reorganisation. Miss Tordoff's role as Assistant Secretary, which had remained vacant since the end of September, was filled in December with the appointment of G. K. Stead ACIS.[702]

The overseas losses which had wreaked havoc with the 1962 profit figures were unsustainable and in January 1963 the Directors discussed the possibility of saving money by amalgamating with other buying brokers or closing one or more branches. The impact of the losses on the Australian operation over the next few years was to be extreme (see p. 203).[703]

In the same month, it was agreed to appoint W. T. B. Jowitt's son, F. T. B. Jowitt, to the Board effective from 23 February. While he was not yet considered experienced enough to run a department, it was important that he was a Director in order to look after the family interests. Also in January, it was announced that the Meanwood fellmongery was only operating for three days a week, a situation which was to continue for some months.[704]

In April 1963, the Managing Director, J. B. Harland, reported that the company faced a possible loss on futures contracts on 23 May. He had used them to hedge against sheepskin price fluctuations and they should have been bought back when the sheepskins were used but this had not happened. With prices rising, the losses were increasing.[705] Due to low sheepskin stocks and high skin prices, the fellmongery continued its three-day week in May. The Board also discussed the charges for efflu-ent from the Meanwood plant, which was felt would become a serious problem in future. Peter Bell was deputed to resolve the futures problem.

In May 1963, Peter Bell, attending the Board meeting for the first time, said that Hollings Mill's production had improved by comparison with that of 1962, mainly due to the improved quality of the raw wool. It was felt that a new bonus system was needed as the basic wage was too low. Revenue for the first six months was £11,000, excluding futures losses but including a loss of £21,000 in the Australian branches.[706]

In July 1963, production at Hollings Mill was up on the previous year due to overtime working. By contrast, production and profitability in the fellmongery had been hit by the three-day week. It was estimated that only the heavy wools were profitable on such a basis. Meanwood had also lost labour and F. T. B. Jowitt was looking into a new payment scheme for pullers. The company was also considering enzyme treatments in the fellmongery, although it was acknowledged that they were still experi-mental. F. T. B. Jowitt was asked to investigate the production and run-ning of Meanwood and report to the Managing Director. F. R. B. Jowitt, who had been too ill to attend Board meetings for some months, retired at the end of July. The Board expressed their gratitude for the valuable work he had done for the company and the very high esteem in which he was held by his fellow Directors.[707]

In October 1963, the output of Hollings Mill was reduced by the need to carry out extensive combing tests on Australian wools prompted by concerns over misbuying (see p. 200).[708] Also in October, two new ventures were considered. Jowitts had developed a trade in scouring and exporting English wools but some English merchants in scoured wools were planning to bypass Jowitts. A discussion was held with Mr A. Hud-son, the Bradford Manager of the English, Scottish & Welsh Wool Grow-ers Association Ltd, with a view to setting up a company which would help to maintain the scouring and export of English wools. It was agreed to investigate the possibility of creating a subsidiary for such a purpose. At the same time, P. J. M. Bell suggested setting up a Mohair Depart-ment, for which there was little competition in Bradford. On the sugges-tion of Ted Bell, it was agreed that a Mr Irvin Sykes (formerly of Jeremiah Ambler & Sons Ltd, mohair spinners) be interviewed as a potential mo-hair buyer.[709] Matters proceeded rapidly and a Mohair Department was set up in November. Mr Sykes would be employed on a three-year service agreement to run both buying and selling operations. Jowitts would provide finance of approximately £100,000. Processing would be placed in the hands of commission combers. Plans for an English wool company had also progressed. Mr Hudson handed in his notice to his former employers and Jowitts started looking for suitable premises for the new subsidiary. A shortlist of three possible names was drawn up: Dales Wools Ltd, Abbey Wools Ltd or Wharfedale Wools Ltd[710] and by the December meeting Dale Wools Ltd was in the process of formation. High skin prices were still restricting the output of the Meanwood fellmongery.[711]

For the 1963 financial year, the company made a loss of about £30,000 on futures, a £28,000 loss on the Australian operation and a small operating loss on the Meanwood fellmongery. Despite this the company made its best profit in a decade, at £126,800 (see Appendix XII), thanks to a number of items such as surplus interest charges and appreciation of pelts. Stock values at the end of the calendar year were high due to a combination of high stocks and equally high prices.[712]

By January 1964, Mr Sykes had joined the company and small quantities of mohair had been bought to produce type samples. Mr Sykes was to visit South Africa in March for the next mohair sales. Although Dales Wools Ltd was still in the process of formation, sorting of English wools had already begun, premises had been obtained and storage faciliities would soon be available.[713] Skin shortages were still hampering production at Meanwood, although they improved in April.[714]

The trading figures for the first seven months of the financial year to May 1964 were excellent, but high stock figures and falling prices meant that the company was facing a stock devaluation of £53,000. English wools had experienced a similar drop in prices and Dales Wools posted a £1,500 loss in its first three months of trading and a further £700 in May and June and similar losses for July and August. The company held mohair valued at £43,000 at the end of May and was experiencing difficulty selling it. Although the company had made its first mohair sale by late July, export reaction to the product was unpromising.[715]

Fluctuations in wool and top prices had proved fatal to the year's trading. In the first part of 1964, wool prices had risen more than top prices, but when wool prices fell later in the year, top prices had followed them down. At the end of the year to 31 October 1964, the company was faced with a loss of approximately £90,000 after writing £108,000 off stock values.[716] £30,000 was subsequently transferred from the reserves to the profit and loss account.[717]

The end of the 1964 financial year had also seen a loss of approximately £10,000 at Dales Wools Ltd. Jacomb Hoare & Co., Wool Brokers Ltd, with whom Jowitts had done joint account sales of about £500,000 during the past year wanted to increase their scope. It had been proposed that the capital of Dales Wools Ltd should be increased to £50,000 and that Jacomb Hoare would then take a 50% interest. Jacomb Hoare would channel their joint account and English wools business through Dales Wools and also transfer a South American wool agency to the company. The Board agreed, subject to satisfactory financial controls.[718]

In November the company was approached by Peter Fawcett, Managing Director of Richard Fawcett & Sons Ltd (a subsidiary of Robert Clough & Co. Ltd), and Cyril Reddihough, Chairman of the publicly-quoted John Reddihough Ltd, with a view to a three-way merger of their businesses. The companies were facing increased competition from Woolcombers Ltd, which had recently acquired several topmaking companies. The Directors in daily attendance did not favour a merger which included Richard Fawcett & Sons Ltd as it was believed to have surplus combing capacity and was reliant upon making physical deliveries onto the London Wool Terminal Market rather than to worsted spinners. They would prefer a merger with Reddihoughs alone, with an eye to further mergers in the future.[719] The boards agreed to commission an independent survey of the three companies and the market for tops to establish the comparative merits of remaining independent or merging. It was felt that if Robt. Jowitt & Sons remained alone, 'drastic economies would have to be made, with branch and departmental closures, and standardisation and reduction of stocks.'[720] Kleinwort Benson was commissioned to undertake the survey.[721] Although there is no indication that the Jowitts were aware of it, Kleinwort Benson was the heir to the financial empire built by their more distant Benson, Braithwaite and Rathbone cousins, including Constantine W. Benson who had guided some of their North American investments in the nineteenth century. Ultimately, the merger was rejected by Robt. Jowitt & Sons on the advice of Peter Bell (who had been delegated to represent the company in the negotiations) because the maximum 35% share offered in the combined company failed to reflect the true value of the Jowitt assets,[722] in particular its two-thirds of tops produced on commission combs. Moreover, the Jowitt board believed the plan only made sense if both the Fawcett and Jowitt plants were closed, with combing being consolidated at the recently refurbished 44-comb Reddihough mill at Batley, whereas the Fawcett directors refused to contemplate the closure of their plant.[723]

The losses of 1964 had rid the Directors of any lingering complacency and if anyone imagined that Robert Benson Jowitt II, so often absent from Board meetings due to illness, was a spent force, they were soon disabused. On 24 March 1965, the second day of a marathon Board meeting, which discussed, amongst other things, the closure of all Australian Branches except Melbourne (see p. 203), he told the Directors that,

> The Company must be run with the interests of all concerned in mind, and rather than cut staff indiscriminately, the Directors should face the problem, or close the Company.
>
> Export and Woollen Departments [which it had been suggested should be merged], although showing a record of losses, served as an entry into the woollen trade, and the Company should not be reduced to a skeleton because of one year's bad results.
>
> The Departmental question would be decided by the Managing Director, who would consider proposals put to him by the Directors in charge of Departments.
> The Chairman stated very strongly that he believed in the Company's future, that any sacrifices to be made must first be made by Directors...[724]

The forceful words were rather undermined by RBJ II's insistence that his own fiefdom, Woollens, should not be merged and further undermined when he added rather lamely that he would take a cut in salary 'if necessary'. It is perhaps unsurprising that his warning had little immediate effect, and the Managing Director reported at the next meeting that he had not yet received any recommendations from Directors in charge of departments. The Chairman reiterated that Directors must work together more for the benefit of the company.[725]

The Chairman continued to emphasise the necessity for change at the May Board meeting, stating that,

> being a private family Company with no outside interests, the Company could not continue with losses. There must be economies and complete

collaboration, but that the present day conditions should not be the basis for any changes.

During the general discussion, it was stated that we had been prepared to amalgamate with other Topmakers, but the discussions had failed; vertical amalgamations were not possible, as Spinners could buy their requirement cheaper from independent Topmakers. It appeared that during the past year synthetics had made an impact to the detriment of wool. The main question was how far should, and could the Company contract, so that on a stable market the Company could break even.[726]

It is puzzling that Robert Benson Jowitt II's rescue plan should be predicated upon a stable market. Both he and the Managing Director, J. B. Harland, despite their long experience of the trade, seem to have expected that wool prices would cease to fluctuate wildly as they had done for hundreds of years. It was a delusion similar to that shared by those who at the end of the twentieth century proclaimed that inflation was dead. James Harland said that,

in the past profits and losses had been made on taking a view on wool, and that *in future the price of wool would have to level out more* and profits be more dependent on turnover and efficiency of conversion.[727]

It was agreed that a scheme should be drawn up to merge the Export and Woollen Departments with the Top Department.[728] By July the amalgamation was under way, although it had been established that the savings from redundancies were small and it was decided to rely on natural wastage. It had been intended to house the combined departments on the second floor but this had been found to be impracticable, so it was decided to move them to the first floor and move the Board Room and Directors to the second floor. The Managing Director explained that the object of amalgamation was to streamline bookkeeping, increase efficiency and eliminate departmental jealousies. Better utilisation of stock, closer liaison and central buying would create savings.[729] That this was overdue was probably proven by the case of a customer who bought a few bales from the Woollen Department and then walked downstairs to the Top Department where he sold them at a profit to the Topmaking Director, Harry Ingham.[730] When Ingham realised what had happened, he complained, but F. T. B. Jowitt responded that the man had shown great initiative and was the sort of person Robt. Jowitt & Sons should be employing.[731] Although some Directors argued that the company's overheads had been contained in recent years, they were told that this was just the beginning and that constant reviews would be made. The Chairman stated that personal preferences must be put to one side and that it was essential to improve efficiency. It would be impossible to pay an Ordinary Dividend this year. If the mood in the Board Room had not been sombre enough already, news of the death of F. R. B. Jowitt prompted the Directors to stand in silent tribute at the October meeting.[732]

In December the Chairman informed the Board that the trading loss for the year to 31 October 1965 had been about £257,000, which he described as disastrous. Together with the loss for the previous year, it meant that more than three quarters of the company's reserves had been wiped out. He stressed that no firm could continue like this and that drastic action must be taken. It was too early to establish how much money the Australian closures (p. 200) would save the company. He outlined some further measures which he thought should be taken before the Annual General Meeting:

(1)   All Directors earning £3,000 a year or more to take a 10% reduction as from 1st January, 1966, until the company makes a profit. Present pension rights to be retained if possible.

(2)   All departments to be scrutinised to see if there should be any redundancies or more efficient methods. If redundancies were necessary, and the employee had served at least 20 years, redundancy pay should be as generous as Australia.

(3)   Consideration of continuing Meanwood in its present form.

(4)   Liquidation of Indian Investment (see p. 215) if at all possible.

(5)   Investment in Maryland (see p. 210) to be liquidated if the new process had not proved successful within the next six months.

(6)   The Company's investment in Dales Wools Ltd to be considered.

(7)   Weekly meeting of working Directors to be held each week at 11 a.m. on Fridays.

(8)   Quarterly meetings of the full Board of Directors.

(9)   The undeveloped land in Australia to be realised.

The Chairman also stressed that he relied on the two younger Directors, F. T. B. Jowitt and P. J. M. Bell to play their part in the changes and that leadership should come from the top and from the Managing Director. F. T. B. Jowitt should consider himself the Chairman's assistant and keep him informed in all matters. There was resistance to the 10% pay cut for Directors and a cut of just 5% was finally agreed.[733] This, as it turned out, was Robert Benson Jowitt II's last Board meeting. Suffering a heart attack while at work, he was rushed to Bradford Royal Infirmary by F. T. B. Jowitt but did not survive.

THE ART OF WOOL BUYING

Judging the tear (top and noil yields) of greasy wool from a quick visual inspection had never been an exact science; in fact, it had never been a science at all but a subjective process which often proved fallible. A steep decline in wool values since 1957 together with increased competition on the sales side had led to the demise of the long-established ±1% allowance on estimated yields compared with combing results. Buyers, working under increasing pressure, were more likely to make mistakes in visually assessing the likely top and noil yield of wool bales prior to auction. A more scientific approach had been adopted in the United States, whereby a core sample was taken from the bale by manually inserting a narrow tube with sharpened tip into the bale. Later, power-driven rotary coring machines were developed which could take a sample from at least 95% of the length of the bale, thus being more representative of the bale's contents. The importance of introducing such techniques was shown by Jowitts' problems in Australia (see p. 200).

In 1961 Wool Testing Services Ltd introduced a UK laboratory for top and noil testing of greasy wool. The UK adopted a mechanical rather than chemical test as used in America but by 1968 an International Wool Textile Organisation (IWTO) chemical test method had been agreed which, apart from allowances for residual ash and alcohol extractables, was essentially the same as the American method. These allowances were soon brought into line with America.

Core-tested results for yield and fibre fineness (micron measurement) enabled combing blends to be made without the necessity of making individual combing tests to confirm yields. Core testing removed the need for the subjective skills of the buyers and topmakers and weakened their claim to what many in the industry considered excessively high salaries.

Pre-sale testing was introduced in the 1973 wool selling season, allowing buyers to have greater confidence in the yields of the wools they were buying. Because the bales no longer had to be open for visual examination, wool handling costs prior to auction were reduced and the bales could be more highly compressed, reducing display and transport costs.

Further developments in wool metrology combined with sale by sample, obviating the display of bales prior to auction, were championed by the International Wool Secretariat's Raw Wool Services Department which was headed by Peter Bell between 1974 and 1989.

AUSTRALIA

With the closure of the Adelaide fellmongery in 1959, the Australian operations had been left to concentrate on wool-buying. There were promising developments, such as firm-offer business from Japan, a joint buying venture with the Japanese firm Gosho & Co., and associate membership of the Sydney Greasy Wool Futures Market. With more countries installing combing equipment, greasy wool buying was seen as the best growth area for Jowitts to target, although the possibility of an Australian combing plant was also investigated. There were also plans to enlarge the Perth office to give extra storage space and install new dumping (i.e. compressing) equipment.[734] More radically, Jowitts invested in an Indian combing plant (see p. 204) in the hope that it would become a major customer for its Australian wool-buying operation.

Unfortunately, nothing went according to plan. In the first place, the concerns about poor buying which had surfaced in the fifties were shown to be fully justified. The Managing Director, J. B. Harland, informed the Board in December 1960 that incorrect buying had caused a serious loss of profit. It had been found that estimated yields for inferior and more heavy conditioned wools had been 6–8% wrong, leading to an over-all error of 2%. He had discussed the matter thoroughly with the Australian Director, Simeon Myers, and the following had been agreed:

Immediately upon his return to Australia Mr. Myers should convene a Meeting with Mr. Gregson, of Perth Office, and Mr. Townend of Melbourne Office, and make them fully cognizant of the position and of their position as Head Buyers, which called for stricter supervision. Regular testing on the basis of 20% of wools purchased would be made at Bradford, to implement these remedial measures, and any underyields would be charged to the relative Buying Office.[735]

Despite this, the company felt obliged to increase wool-buyers' salaries by 10% and those of office staff by 5%. There had been no pay review for Australian staff since 1957 and the cost of living had increased substantially.[736]

Any hopes that misbuying had been eradicated were shattered in September 1961 when J. B. Harland reported that core testing had shown misbuying of wools from Perth[737] to the extent of 4–6% over-all, with variations of as much as 10%. This was confirmed by the visual estimates already made by Geoffrey Harland and Harry Ingham. Robert Benson Jowitt II reported that he had reprimanded all the Australian buying staff, particularly those in Fremantle, during his recent visit. He added that no explanations of the differences had been made. The Board felt strongly that Gregson, as Fremantle Manager, was culpable. It was also felt that Simeon Myers was not being firm enough in his supervision. It was agreed that,

> Mr. Myers be instructed to attend the next Fremantle Sale, and apparently in the course of his normal duties make a spot check of the Buyers and their Buying and report back immediately to Head Office: that a substantial portion of the wools received be combed as a test in Hollings Mill in order that the Bradford estimates might be confirmed, and the core-test yields verified, after which an unbiased decision could be made as to the retention or otherwise of the culpable buying staff at Fremantle. The Chairman undertook to write to Mr. Myers immediately on receipt of the salient facts from Mr. Geoffrey Harland.[738]

The situation was not improved by Simeon Myers' aversion to flying between selling centres. Although he accepted that there had been bad buying, he did not suggest any solution. Geoffrey Harland said that Gregson had been informed of the problem a year ago. Harry Ingham thought that Ian Yeoward should be transferred to Perth, while Gregson was transferred to Sydney. However, the Board felt that it would be difficult to implement changes during the buying season, and J. B. Harland thought that any decision should await the combing trials in Bradford.[739] Meanwhile, underyield claims during the last year gave a detailed picture, indicating slightly better yields, of the wrong quality, with too much of the 60's as against 64's. It appeared, said Geoffrey Harland, that all the buyers at Fremantle were at fault and that the Chairman's warnings during his visit to Australia had been ignored.[740]

Gosho Jowitt, the joint buying operation which had seemed so promising, came to nothing when Gosho decided to move all its northern centres business and all its southern centres firm offer business on the basis of guaranteed yields to the French company Dewavrin in 1961.[741] Given the underyield problems which Robt. Jowitt & Sons was experiencing and Myers' strong aversion to the Japanese, this was hardly surprising.

Geoffrey Harland circulated the results of the combing tests at the January 1962 Board meeting. They simply confirmed all the previous assessments and it was decided to offer Ian Yeoward (a cousin of the Harlands) the managership of the Fremantle office and move Gregson, not to Sydney but to Melbourne.[742] Gregson was later told that there was no alternative to this move,[743] although the company later relented to the extent that he was given a choice of going to either Melbourne or Sydney.[744] Ian Yeoward moved from Melbourne to Fremantle, while D. R. H. Yeoward moved from Adelaide to Melbourne.

Geoffrey Harland reported to the Board in July 1962 that, following Ian Yeoward's move, the Fremantle office was very efficient, as were the Melbourne office under Jack Townend and Hamilton and the Sydney office under Myers. D. R. H. Yeoward had yet to transfer to Melbourne. Adelaide was doing little business and it was decided to dispense with the services of George Rushby, their long-serving buyer there. He would receive the full benefits of the superannuation scheme and be invited to act as their agent. As the rent was cheap, it was decided to keep the Adelaide office for use during the wool sales.[745] Even after these changes, the Australian branches were showing heavy losses. In January 1963, J. B. Harland warned the Board that Australian wool prices were too high compared to Bradford prices and it was felt that one or more branches might have to close. Merger with another buying organisation was also considered.[746] In February 1963, Geoffrey Harland suggested closing the Sydney office and buying through brokers at ½% commission. However, when Kreglingers were approached about the possibility of acting for Jowitts, their charge structure made the saving too small to be worth considering, especially if savings could be made in the Sydney office. The Managing Director was requested to write to Simeon Myers asking him to make savings of the order of £A1,500 in the Sydney office. It was thought that this could be achieved by reducing travelling and dispensing with one secretary.[747]

Further concerns were raised about the quality of Australian buying in 1964 when it was said that two recent Adelaide shipments had been very bad and some German combing results and Indian shipments were also wrong.[748]

In 1965, Peter Bell told the Board that the company had a minimum requirement of 45,000 bales per annum and an analysis of purchases in recent years indicated that only 13,000 of those would be bought through the company's Australian branches. The Chairman said that with the high bank rate it was clear that the Australian branches employed on indent buying could not survive. The Board therefore took the momentous decision to close all Australian branches other than Melbourne, where a senior buyer, two under-buyers and the necessary support staff would be retained. The Melbourne office would also be used as an agency for 'any business which might be profitable'. It was proposed to sell the company's land at Melville, Perth, and use the proceeds to purchase new offices in Melbourne.[749] This land was eventually sold in 1966 to Marada Compañía Naviera SA for $A88,000 (= £A44,000 – the Australian Dollar was introduced on 14 February 1966).[750] As always, the changes were considered in the light not only of finance but of the people involved. The Chairman, R. B. Jowitt II, speaking about the whole company, not just the Australian branches,

> addressed the Board in very strong terms on their responsibility as Directors for the affairs of the Company. He stated that any changes that had to be made would be made without undue hardship to individuals, and that the Pension Scheme must be used to prevent such hardship.[751]

Simeon Myers suggested that the rental of £A4,300 which could be obtained for the Fremantle office was less than the income from sorting and dumping and that it should therefore be kept open. However, Geoffrey Harland argued that purchases of Australian wool would be much reduced in future and the Fremantle office would then become uneconomic. It was agreed to proceed with the policy of closing all branches other than Melbourne.[752]

The Sydney and Fremantle offices duly closed on 31 July 1965, although Ian Yeoward remained as buyer in Fremantle. The Fremantle property was placed for sale with Joseph, Charles, Learmouth Duffy & Co.[753] but was subsequently leased for three years to Associated Wool Dumpers Pty Ltd for £A4,250 per annum,[754] renewed for a further three years in 1969 at $A12,250.08 p.a.[755] Six staff were declared redundant, with long service leave and redundancy pay totalling a meagre £A7,089 on top of the money due under the company's retirement schemes.[756] It was learnt that Miss Grady, who had been with the company for forty-three years, had not been enrolled in the pension scheme due to its high cost and the Board decided to give her an additional £A1,200. Simeon Myers left at the end of October.[757] An unforeseen consequence of the company's contraction in Australia was that the Bank of Adelaide withdrew its credit facility. Since the closure of the branches increased the company's financial security, this was a rather foolish action which drew the directors' indignation.[758] They were able to indulge in *Schadenfreude* when the Bank of Adelaide had to be taken over by the ANZ Bank in 1979 when a subsidiary sustained heavy losses on unsecured loans.

In January 1966, the former Australian Director, A. A. Gibson, suggested that the company became involved in up-country buying but the Board considered that the limited facilities made this unfeasible.[759] In June of that year, J. B.Harland told the Board that it was intended to spend a minimum of £8,000 per week on Australian wool to ensure a stable minimum annual income for the Melbourne branch.[760]

INDIA

With the company's traditional markets shrinking, emerging markets such as India became very important. The partition of India at independence in 1947 had left much of the country's traditional textile-producing area in Pakistan and the Indian government was anxious to reduce its dependence on foreign manufacturing, including topmaking. As early as 1959, the company had been approached by Parkash Bros to see if it wished to collaborate on installing a combing plant in India but the Board declined the offer.[761] However, economic realities soon made investment in Indian combing seem attractive.

In March 1961 the Board discussed the likelihood that India would impose import duty on tops. The Managing Director identified three tactics for retaining a share of the Indian market:

(a) The shipment to India of Matchings purchased in Bradford for combing in India.

(b) The purchase through our Australian Branches of core-bore tested Greasy Matchings to be shipped to India for combing.

(c) The shipment of Australian and/or other types of Greasy or skinwools for commission sorting in Calcutta, prior to combing in India.[762]

It was decided to run a trial of 200 bales of Australian core-tested wool, sending half for combing in Bradford and the other half for combing in India.[763] The trial was revised, expanding the quantity to 400 bales to produce a wider ranger of tops.[764]

In July 1961, J. B. Harland told the Board that the Indian Government was interested in the development of a complete textile industry. To this end, Modella Woollen Mills of Bombay had been authorised to construct both a combing plant and a woollen mill of 2,000 spindles in Chandigarh. This city was the new capital of Punjab built to replace Lahore which, following Partition, was now in Pakistan.[765] In August 1961, Mr Brij Mohan Grover, the Managing Director of Modella Woollens Ltd, met the Chairman, Robert Benson Jowitt II, asking the company to co-invest in the Indian combing plant. He was anxious to obtain a letter expressing, in general terms, the company's interest. This would allow him to approach the Indian Government whose interest in encouraging foreign investment would facilitate the granting of the necessary permits. He was also keen that Robt. Jowitt & Sons' name should appear on the prospectus for floatation.[766]

Since this proposal seemed to solve the company's problems in the Indian market, the Board received it very favourably and it was unanimously resolved to give Mr Grover the letter, provided that (1) Robt. Jowitt & Sons was granted sole rights to buy wool for import into India for the new company for a period of seven years, (2) that it be granted a place on the Board of Modella Woollens and (3) that the figures presented to them by Mr Grover were confirmed. The Chairman also felt that it was essential for a company to have a 'responsible person' in the Indian mill.[767] Negotiations went well and in November J. B. Harland and

Harry Ingham visited India. Harland reported back in December that the mill would probably use more Indian wool than had originally been envisaged and that there would be more Schlumberger (i.e. French) combs and fewer Noble combs, Schlumberger combs being more flexible and more internationally acceptable. The Board were told that Modella might not require the company to take up their shares for two years, delaying the return on their investment, but this did not seem to cause any concern. Nor did the requirement to reassure the Indian Government of the commitment to collaboration and transfer of know-how. A partner of Crawford Bayley & Co. of Bombay was appointed to act as the company's attorney with regard to the prospectus and Harland and Ingham were authorised to sign an agreement with Kanshi Ram Kidar Nath regarding any future sale of Modella shares.[768]

One of the two Managers of Hollings Mill, Arthur Johns, was sent out to Chandigarh to supervise the construction and fitting out of the Indian combing plant. He reported back that he had authorised overtime as work on the building was proceeding very slowly.[769] Johns remained manager of the Indian mill for several years.

In consultation with its solicitors, the company sent Modella a revised draft of their agreement in May 1962.[770] However, it was the unrevised draft agreement which was approved at an EGM of Modella Woollens Ltd. J. B. Harland explained that he had been unable to reach agreement with Mr Grover about some essential terms of the collaboration and it had been decided to postpone the passing of the final agreement until a meeting in August. The main problem was that Modella was unable to grant Robt. Jowitt & Sons sole supplier's rights. This was a major blow since such an arrangement was central to the project's viability for Jowitts. The other sticking point was that Modella would not agree to pay the company 3% commission on purchases.[771] With the company initially insisting that 3% was the minimum commission which was acceptable and Modella insisting on 2¼%, they eventually compromised on 2½%.[772] Two agreements were signed by the parties on 12 December 1962 but these were later cancelled and replaced by a new one in March 1963.[773] The two recently-appointed directors, F. T. B. Jowitt and P. J. M. Bell, expressed their concern that the agreement was unrealistic and the proposed investment of £110,000 was not justified by the likely return to Robt. Jowitt & Sons. The Chairman countered that the new directors had not been appointed to question their seniors' decisions. The dissenting directors' opinion that the agreement should not be signed was rejected, but events proved their stance to have been correct.[774]

By February 1963 the new combing mill was nearing completion and it was expected to open in late March or early April. The mill was indeed open in April, with seven combs running on a two-shift basis. Spares were in short supply but the company had a licence to import £14,000-worth.[775] Unfortunately, in May production was hit by constant breakdowns. The 1962–3 Sino–Indian War had caused the Indian Government to divert its foreign exchange reserves leading to drastic import restrictions. As a result, spares were difficult to obtain and the mill was having to work with only Indian wool.

Modella indicated in February that if Jowitts wished to participate in the new rights issue, it would cost them about £40,000 which was more than the company was willing to commit. The Board considered taking part in the issue and then offloading their shares on the Indian market but in the end difficulties in the Indian money market and barriers placed in the way of foreign participation dissuaded them from doing so.[776] However, concluding that they were obliged to do so by the terms of their agreement, the company reversed its decision and took up shares in Modella at a cost of £100,148, requiring an increased overdraft facility to do so.[777]

In September 1963, despite teething problems and the fact that no import licence for wool had been granted by the Indian Government, it was decided to expand the mill's capacity. It was hoped that the licence would be forthcoming in a few months, but it had still not been granted by February 1964. Without this, Jowitts could not make any commission and it was suggested that it might be possible to vary the agreement so that the company received a commission on combing instead of the supply of foreign wools.[778] Mr C. C. Desai, Chairman of Modella, visited Jowitts in June 1964 and explained that the Government was trying to conserve foreign exchange (which had been diverted to military spending due to the Chinese attacks) and was also being pressured to restrict wool imports, despite advice that Indian wool was unsuitable for combing. It was anticipated that the exchange situation would improve in the short term. He said that it should be possible to obtain customs clearance permits for imported wool but conceded that these would be of little value without an import licence. R. B. Jowitt II pointed out that Robt. Jowitt & Sons had entered into the agreement in order to enhance the buying potential of its Australian branches, something which could not be achieved while the import restrictions remained in place. He was concerned that what the Indian Government considered short-term might be five or seven years, and the company was now some way into its a seven-year contract.[779]

It was only at this meeting in June 1964 that Jowitts learnt that Modella, as well as expanding the combing, was intending to create a vertically-integrated business on the Chandigarh site, combining combing, spinning and weaving. The Indian directors claimed such a business, which could earn foreign exchange from the export of cloth, would be exempt from the exchange controls which were inhibiting the company's growth. Peter Bell pointed out that the enlarged combing capacity of 5 million lb per annum would far outstrip the possible spinning and weaving capacity. Mr Desai said that the extra combing plant could be moved to Nepal, which had no exchange controls. In any case, it seemed that the plan for an integrated site might never come to fruition as it was opposed by the Indian Wool Federation, of which Mr Grover was Chairman. On a more positive note, Mr Desai felt sure that Modella would be open to a royalty on production. This was not very re-assuring to the Board since the company's involvement with Modella had been motivated by a desire to create a new market for the Australian operation, something which this royalty would not provide.[780]

There were still no customs clearance permits or import licences in July 1964 and it was decided to raise the question of a production royalty with Mr Grover when he visited the UK.[781] The CCPs had been received by September but Modella had been refused import licences. By November the Board had decided that they must seek a royalty on production.[782] Modella applied to the Indian Government for permission to pay technical fees of 1d per kilo of top and noil production of the Schlumberger plant, which, it was argued, was the highest rate which the Government might entertain. However, the Jowitt Board were not happy with this figure which had been arrived at without consultation. It was estimated that this would yield £3,600 p.a. nett of Indian tax, compared with the £14,000 which had been anticipated from the Australian buying commission. The Board insisted that the technical fees should be at the rate of 1½d per kilo and should include the Noble as well as Schlumberger combing. Should that not be acceptable, Robt. Jowitt & Sons wished to sell their shares in Modella in accordance with the agreement of 21 November 1961 between the company and Kanshi Ram Kidar Nath.[783] However, J. B. Harland reported that the Chief Controller of Capital Issues of the Government of India had told him that the shares would have to be sold for Rupees and that it would be extremely difficult to get permission to repatriate the Sterling equivalent.[784] In October, the Indian Government gave permission for the technical fees for French combing only, as this used Indian wool. The Government also forced Modella to cancel the renewal of contracts for two English technicians.[785]

In March 1965, F. T. B. Jowitt was appointed second director of Modella, to replace Douglas Penney Barker. His appointment had been actively sought by Modella's Chairman, C. C. Desai, and Managing Director, Brij Mohan Grover who believed that the young man would be easy to bend to their will. They were presumably unaware of his vocal opposition to the partnership in 1962. J. B. Harland reported at the end of that year that the Indian economy was in a very poor state, meaning that Jowitts could not sell Modella shares at a reasonable price and, at the same time, import restrictions would continue to prevent the purchase of Australian wool. Indeed import restrictions were now affecting the supply of raw materials to Modella's shoddy plant. Even more seriously, annual Indian wool production suitable for combing was estimated at 12 million lb, while the Modella French combing plant alone could handle 5 million lb. It was said that no more import licences would be available for wool and the only way to import it would be through export promotion licences.[786]

SOUTH AFRICA

By 1960 the South African operation's main income was from sheepskins. The company faced fierce competition at times from French buyers, but benefitted from skilled buyers, drawing on a knowledge of yields built up over the previous forty years.[787] When the company received a tentative invitation to invest in the fellmongers W. Lane & Co. (Natal) Ltd in 1961, it seemed like a natural fit.[788] After a report by Peat, Marwick, Mitchell & Co. and some haggling over price,[789] the company bought a 51% interest at a cost of R34,425 (7,650 R2 shares at R4.50 each) in November 1961, agreeing at the same time to guarantee Lane's overdraft up to R30,000.[790] Fellmongering costs were very low in South Africa and a report by J. Smith indicated that the Lane interest should be 'very beneficial'.[791] Later the company tried, without success, to buy a further 700 shares in Lanes from a Mr Howells.[792]

In 1963, the company was invited to take a third share of another South African fellmongery, Maryland Wools (Pty) Ltd. Although Maryland had been incurring losses, it was decided to take up the offer.[793]

Frederick Thomas Benson Jowitt

From now the pound abroad is worth 14 per cent or so less in terms of other currencies. It does not mean, of course that the pound here in Britain, in your pocket or purse or in your bank, has been devalued.

Harold Wilson, 19 November 1967

Robert Benson Jowitt II suffered a heart attack at work on 11 January 1966 and died while he was being driven to Bradford Royal Infirmary by his nephew F. T. B. Jowitt. At the Board meeting on 26 January 1966, J. B. Harland spoke of the great loss suffered by the company, adding that everyone had lost a great friend and colleague. The meeting stood in silent tribute to their late Chairman. Mr Harland went on to say that it was well known that F. T. B. Jowitt was being groomed as the Chairman's successor and, while his time had come rather sooner than expected, he had great pleasure in proposing F. T. B. Jowitt as Chairman. This proposal was seconded by R. McC. B. Jowitt and approved unanimously.[794]

It was a big step for a young man who three years before had been considered too inexperienced to run a department (p. 192), and the task was all the more daunting for the figures for the financial year to 3 October 1965 which had been announced in December and were confirmed in the January meeting. Nor was it helped by the news that Barclays Bank was cutting unsecured lending to the topmaking section to £450,000.[795]

The growing competition from man-made fibres was raised by J. A. Jowitt at the June meeting, but he was told by Peter Bell that the company was well aware of the potential demand for anti-shrink wool tops.[796] As it happened, the company was to become a specialist in this area.

Far from operating in a stable market as R. B. Jowitt II and J. B. Harland had apparently anticipated, the company continued to suffer from serious fluctuations, finishing the year with a stock write-down of about £38,000. Worse still, the losses of previous years had forced the company to finance more of its buying with bank loans causing a massive increase in interest charges to £87,000. The Meanwood fellmongery was losing money, too. It was paying full wages while working on half time. Only short-wooled sheepskins were available at that moment and it was doubted whether there was a demand for the resultant skin-wool. It was agreed in January 1967, Peter Bell and D. P. Barker dissenting, that the

211

Highbury works, including the Meanwood fellmongery, should close once stocks had been run down.[797] There had been talk of selling the site to a television consortium which included the *Yorkshire Post* but the broadcasting licence went to a rival group. In the end the land, buildings, plant and machinery were sold to F. M. C. (Products) Ltd, a subsidiary of the Fatstock Marketing Corporation. F. M. C., unlike Jowitts, had access to a ready supply of English sheepskins.[798] The company had hoped for £125,000 but only received £70,000 because the land only had planning permission for a fellmongery and no change of use was allowed, a surprising decision considering the noxious nature of fellmongery. As a result, F. M. C. was the only bidder.[799]

The Chairman commented that 1966 had been a difficult year and that he could not see the company making a profit in the year to 31 October 1967 'in its current form'. He felt that Jowitts had not done as well as some of its competitors and he warned that there were indications that the trade was contracting and that it would be difficult to make large profits in future. It was this realisation which was to lead the company to diversify away from its traditional activities. An advertisement was placed in the *Financial Times* in the late 1960s stating that a company with offices in South Africa and Australia was seeking opportunities for exporting from the UK. One response was a company selling 'natural food'. Testing was left to Tom Sykes, the South African Director, who decided that the food was inedible and threw it into his garden where it grew into a healthy crop of cannabis plants.[800]

Another diversification project was an investment in Scout Alarms Ltd,[801] a company run by Prince Puzyna, a man believed to work for the Secret Intelligence Service. The alarms looked a promising investment; they were relatively cheap and had been approved for use in churches by the Ecclesiastical Insurance Company.

One less radical diversification, Dales Wools Ltd, trading in English wools, was not looking very promising. It had continued to make losses and it was decided to split it in two with Jowitts taking the British wool part and moving it to Meanwood. Jowitts also bought 50% of another English wool company, John S. Dyer Ltd, and merged it with its British wool interests. The concentration on British wool was expected to make managing the operation easier and the move to Meanwood would reduce overheads and gain export advantage. Jacomb Hoare would take complete control of the rest of Dale Wools. This re-organisation was hampered by unreliable stock figures, for which Mr. A Hudson was dismissed.[802]

Given the serious state of the company's finances at the beginning of 1967, the executive Directors, with the exception of Geoffrey Harland, agreed to waive their entitlement to 7½% of the profits (after payment of the preference dividend but before Corporation Tax).[803]

J. B. Harland, who had been Managing Director since the end of 1959, resigned from the post in March 1967, while remaining on the Board, and was replaced by Peter Bell.

Despite all the efforts made by the Directors and employees, the company still ended the 1967 financial year with another substantial loss (Appendix X). Recent initiatives had disappointed: mohair sales had failed to materialise in quantity and Scout Alarms, to which Jowitts had lent money, having failed to attract the expected ecclesiastical client base, was facing liquidation. Jowitts intended to sell the fifty £1 shares back to Puzyna at par as soon as the debt had been repaid.[804]

Sterling was devalued on 19 November 1967 and, for once, the Directors might have wished that the company's stocks were higher. By the end of January 1968, Hollings Mill had enough stock to last three months, but, with high interest rates and a low exchange rate, it was going to be expensive to replenish stocks, while top prices had remained relatively static. The pound in Jowitts' pocket had certainly been devalued. Mohair continued to perform badly and it was decided to make Mr I. Sykes, the department's manager, redundant from the end of January. It was thought that the department could be run by the remaining personnel. Perhaps surprisingly, the Directors increased their own salaries as well as those of other staff.[805]

The figures for the year ending 31 October 1968 saw a substantial improvement, with a profit of £111,670 before payment of the Preference Dividend amounting to £12,000. However, the Managing Director warned that the company had to improve its nett current asset position over the next few years. The Chairman added that Scout Alarms' debt now appeared to be completely worthless and full allowance had been made for this in the accounts.

Given the state of the company's finances, it was not possible to pay a dividend on the ordinary shares for 1968. The executive Directors agreed to reduce their commission from 7½% to 4½%. After some misunderstanding amongst directors, it was explained that this reduction was meant to apply to the year to 31 October 1968 only.[806]

By 1968, the very core of the company's profitability, the combing operation at Hollings Mill, was underperforming, though the exact reason for this was unclear and in March 1969 it was decided to install two 10-channel production monitoring units on the cards.[807]

The combing performance of Hollings Mill continued to be disappointing, due to in part to poor quality wool. A tight stock policy combined with dock and hauliers strikes as well as reduced top sales led to reduced running hours so that the overheads were borne by a smaller output. To make matters worse, interest rates were high and noils were making much lower prices; claims from customers for poor quality tops also had to be met. The Managing Director emphasised that over a ten year period, mill production had increased despite a reduced workforce, but Geoffrey Harland, the Topmaker, questioned whether this had been at the expense of quality (attack being the best form of defence). While the UK combing operation was causing concern, the company was in the process of setting up an Italian company, Jowitt Italiana S. r. l., to oversee the production of tops combed by continental commission combers.[808] F. T. B. Jowitt was created the grand-sounding Il Presidente of the new concern which, in reality, was merely a device to regularise Jowitts' tax affairs in Italy.

Despite the many problems it was facing, the company managed to turn in a trading profit for 1969 of £41,306, after foreign taxation. In addition, there was a surplus of £20,518 on the sale of shares in W. Lane & Co. (Natal) Ltd (see p. 217).[809]

SHRINK-TREATMENT OF WOOL

A well-known disadvantage of wool products is their propensity to shrink when washed. This is the result of the 'Directional Friction Effect' (DFE) due to the interlocking of the wool fibres' natural scale structure. This awkward characteristic of wool became increasingly significant with the growing availability of domestic washing machines and competition from synthetic fibres which were intrinsically shrink-resistant.

Research to overcome this disadvantage was initiated by the Commonwealth Scientific and Industrial Research Organisation (CSIRO) in Australia and the International Wool Sectretariat's (IWS) research centre in Ilkley under Dr J. R. McPhee. Robt. Jowitt & Sons began trials into the Fi-chlor (chlorine-only) process in 1968. These culminated in the Fi-chlor/Hercosett process (which combined chlorine treatment with a synthetic resin coating). The resultant wool tops, with significantly reduced DFE, were later marketed under the 'Superwash' name. Robt. Jowitt & Sons began trials with shrink treatment in 1968. Development was aided by the good working relationship between Peter Bell, Dr McPhee and his very able colleagues at Ilkley.

Initially trials were carried out at Meanwood using a backwash of pre-war design transferred from Hollings Mill, where two new Petrie McNaught backwashes had been installed. However, it became clear that variable drives on the new Petrie McNaughts were essential to control the top slivers and it became necessary to transfer the processing trials to Hollings Mill.[810] Initially, the only hint of this activity in the minutes of Board meetings was a reference to the purchase of a new padder 'for the Shrink Treatment'.[811]

INDIA

A fall in the Rupee meant that by June 1966 the company's investment and its technical fees were devalued by 36.5%.[812] In the following January, Jowitts wrote down the book value of their Indian investment from £110,000 to £85,149.[813]

The absence of import licences was still a problem in March 1968 but Modella had enough commission combing of Indian wool to keep the plant busy until July,[814] and in October it was announced that Modella were likely to make a profit of 18% on asset value. However, the Board of Robt. Jowitt & Sons suspected irregularities. The Jowitt directors complained that B. M. Grover used Modella to service his private companies, but their request that J. B. Harland's Alternative Director, Mr A. K. Ghosh of Robson Morrow, be permitted to investigate was refused.[815]

After skilful negotiation by the Chairman, Robt. Jowitt & Sons finally sold its shares in Modella in 1970 for £66,319 and was able to repatriate the assets. However, the combing royalty continued to the end of the agreement term and J. B. Harland and F. T. B. Jowitt remained Directors of Modella.[816]

The story remains an object lesson in how governments cannot easily promote foreign investment while maintaining protectionist measures.

Although Jowitts lost money on the venture and the Indian economy benefitted very little from it, the Indian military obtained a number of ageing MiG fighters in return for the resultant wool.[817]

AUSTRALIA

The Australian operation was working well enough at the beginning of 1967 for the Board to sanction increases in Jack Townend's and Ian Yeoward's salaries and to authorise Mr Townend to engage a second buyer at £1,750 per annum to replace Mr Davey who had given his notice.[818] In January 1968, it was announced that Ian Yeoward had resigned and that the Fremantle branch would therefore be closing.[819] The property at James Street, Fremantle, was then leased to Associated Wool Dumpers Pty Ltd[820]. However, the company maintained a presence in Western Australia by appointing G. T. Sadler & G. A. Kennady, Solicitors and Company's Agents, to oversee their interests there.[821]

After all the turmoil of the last few years, it might have been hoped that the small remnant of the company on Australian soil might at last be efficient and profitable. Unfortunately, this was not the case. The company lost money due to a failure of the Melbourne office to take up foreign exchange contracts. This failure was attributed to a Mr Hamilton who left the company in October 1968.[822]

A familiar, and unwelcome, problem re-appeared when Australian losses in the first four months of the 1969 financial year amounted to approximately £4,400, mainly due to underyield claims on wool bought by the company's own buyers.[823] In order to restrict losses in Melbourne, the long-serving Jack Townend was given notice to terminate his employment at the end of October that year, when he was replaced by the younger Brian Downs who was sent out from Bradford at a salary of £2,000 per annum.[824] Downs was instructed to use his experience of the Topmaking Department to meet specific top specifications rather than by wool type. He was also advised to take advantage of core tests for yield and fineness, an approach which was not supported by Geoffrey Harland as it threatened to demystify the craft of topmaking.

SOUTH AFRICA

The South African branch had long suffered from several problems beyond the company's control. Competition for sheepskins, particularly from the French, was fierce, and exchange controls imposed by both the

South African and British governments hampered everyday commercial activity. The Port Elizabeth Office was closed in 1966, with its work being handled from Durban.[825] D. B. Sykes died in 1968 and his place as South African Director was taken by his son T. P. Sykes. By 1969, the South African branch's sole income was from commission buying of sheepskins for the company's agent, Salvisberg et Cie., in Mazamet.[826]

The investment in W. Lane & Co. (Natal) Ltd looked as though it might prove a success. J. B. Harland had told the Board in December 1965 that he was very impressed with both the company and the Managing Director, Mr Williams.[827] However, in late 1968 it was reported that Mr Williams was in hospital and Lanes was likely to make a loss. With mounting losses at home, F. T. B. Jowitt flew to South Africa in 1969 to arrange the sale of Lanes to O. S. Blenkinsop (Pty) Ltd for R7.93 per share, netting Jowitts £17,000. The Durban property belonging to Lanes, which had a book value of R35,374 less R25,000 mortgage, was excluded from the sale and was transferred to a new company, Sunbridge Properties (Pty) Ltd, the shares in which were distributed pro rata to the existing Lanes shareholders, giving Jowitts 51%. As the South African government had blocked the repatriation of the proceeds of the Lanes sales, he used the money to buy out the minority shareholders in Sunbridge Properties.[828]

The 1963 investment in Maryland Wools Ltd was already looking doubtful by 1965. In October the Board was told that it had been impossible to pull any wool there for the last six months due to the high stocks of 'velhair' (i.e. skin wool). It also looked as though it would be too expensive to fellmonger mohair there. It was therefore felt that the company should try to sell its shareholding as soon as it could without undue loss.[829] However, by December it was reported that the other interested parties believed that a new method of processing the skins at higher temperature would work. The company agreed to wait another six months to see what happened.

The Way Forward

A breakthrough in the development of wool garments, which can be machine washed was claimed yesterday by the International Wool Secretariat.

The IWS said that by the autumn of this year some two million garments incorporating its Superwash Woolmark, will be sold in retail outlets throughout the country.

The Times, 29 March 1972

Although 1969 had produced a profit, it was a very small one and it was clear that the company could not continue unchanged for much longer. The figures for the first four months of the 1970 financial year did nothing to reassure the Directors, showing a loss of £11,500 compared with a profit of £21,000 for the same period of 1969. Profits on sales showed a reduction of £21,200 and Hollings Mill's profit was down £13,800 on reduced output. Deliveries and sales had been below the optimum level required to cover overheads. Declining top sales in the domestic market meant that, while Merino top sales were satisfactory as a percentage of the available sales, they were insufficient to maintain Hollings Mill at full capacity. The Managing Director, Peter Bell, warned that tight credit and a steady decline in wool values meant that substantial sales over the next few months appeared to be unlikely, despite the fact that the April/May period was usually a good one for the company. The company had managed to contain losses on raw wool stocks, but its stock of finished tops was too high. The company was currently trading at a loss and was facing increased haulage costs and a 7½% pay increase to all operatives.[830]

Fresh from a three-month Senior Management course, the Chairman, Mr F. T. B. Jowitt, stressed the need for detailed forward strategic planning:

He acknowledged the difficulty in forecasting and planning for a small company in a hazardous trade, but considered that a time span of at least three years should be feasible when compared with the ten and even twenty year spans being considered by larger groups, although not in the Wool Industry.

He considered that the primary function of the Board was to innovate, plan, monitor, and adjust, rather than to concern itself with the consideration and review of the Company's past activities.

The main stress of the late 50's and early 60's had laid on merely preserving the Company and its present activity to the detriment of possible diversification or complete alteration of the Company's activities. The past few years, of necessity had been devoted to preventing a financial disaster, and now, whilst solemn incantations of faith in the future of the Company were no doubt comforting to those making them, they did not forward the Company's course.

The primary consideration to be taken into account when planning the Company's future should be an adequate and growing return to shareholders, and secure and well-rewarded employment for the Company's employees.

In the Chairman's view, woolcombing in Britain satisfied neither of these fundamental imperatives, and accordingly alternative investment projects must be investigated with some urgency.[831]

He said that some carefully devised schemes should be developed prior to the Annual General Meeting. The alternatives to diversification were merger or liquidation. Neither horizontal nor vertical mergers looked very promising and liquidation was 'an admission that the Management had insufficient talent, imagination, or business acumen to do anything else'. He added that he thought that the Board would be better fitted to face these challenges if it were expanded to include those from other industries or disciplines.

An informal meeting of the Directors in daily attendance was later held to discuss redundancies of four staff and a proposal to reduce Directors' remuneration by £1,000 each. A further, formal, meeting was held to discuss these matters on 8 June 1970 at which it was said that Geoffrey Harland (who had been the only Director to refuse to waive his commission in 1967) 'had taken violent exception to these proposals both during the earlier meeting and subsequent to it'. Doubts were expressed as to whether he would be prepared to accept the proposals or would resign but Mr Harland said that he was not considering resignation. He argued that the reductions in Directors' salaries was excessive while the list of redundancies did not go nearly far enough and should include at least two more names, J. Stocker and K. Betts, both expert salesmen under F. T. B. Jowitt. It was yet another reminder of the poisonous split in the boardroom.

The Chairman, supported by the Managing Director, countered that these proposals represented the very minimum by which overheads could be reduced. While he realised that salary reductions in a time of high inflation could cause hardship, he felt that when declaring staff members redundant, the Directors should make, and be seen to make, sacrifices themselves. The Board agreed the reduction in Directors' salaries and authorised the Managing Director to issue redundancy notices to the four people whose names had been placed before them.[832]

Recognising that the best management practices were imperative, the Chairman sent himself on a three-month course at Manchester Business School to acquaint himself with the latest management techniques and to assess his own skills. It was a valuable experience which was to pay dividends in the demanding times ahead.

At a full Board meeting in July 1970, the Chairman spoke of the 'disturbing degree of disagreement between the executive directors and G. Harland'. Although the proposals of the June meeting had been circulated and agreed, he felt that the Board should review these discussions 'to dispel finally any lingering suspicions or disharmony between Members'. The Managing Director stated that he could not continue in office unless he received 'the fullest and most rancour-free co-operation from all Members of the Management.' The Chairman terminated this discussion by asserting that he and the vast majority of shareholders gave their unqualified support to Mr Bell 'who was discharging his duties with great skill and devotion despite the perplexing and daunting trade climate'. He insisted that all members of the Board worked constructively with Mr Bell.[833]

In line with his suggestion in April for adding Directors with experience outside the Wool industry, the Chairman proposed two new members of the Board, Mr W. J. B. Jowitt and Mr C. M. Leveson-Gower. W. J. B. Jowitt was a partner in Bischoff & Co., solicitors and, as the Chairman's brother and a shareholder, was known to all the Directors. He was duly elected. Mr Leveson-Gower, the very able Managing Director of Spooners, a food processing machinery company, was unknown to the other Directors and it was decided to invite him to lunch so that they could decide. He was elected to the Board in December 1970.[834]

The Managing Director's father, Ted Bell, had been in charge of the Top Department for many years and had been with the company for fifty-two years, starting as an office boy at the age of fifteen. It was agreed that, although he was due to retire at the end of 1971, he would be invited to remain as a consultant at £250 per annum on top of his pension. On Ted Bell's recommendation, his position as Manager was taken by David H. Foster[835] who had joined the firm from Richard Fawcett & Sons Ltd where he had been a top salesman.

In September 1974 two new directors were added, Richard Heaton (who was married to R. B. Jowitt II's daughter Patricia) and David Foster. Richard Heaton, a land owner and successful sheep farmer in Wales, was a clever man with an original mind while David Foster had proved himself as Manager of the Top Department. Sadly, Peter Bell tendered his resignation at the same time. He felt he had achieved all he could for the company and had been appointed the first Director of Raw Wool Services at the International Wool Secretariat. The Chairman expressed his regret, both for himself and the company. It was decided that 'owing to the limited scope of the company's current activities,' it was not necessary to appoint a new Managing Director. Rather, the company would be run by a committee of the executive directors.[836] While it may have been true that the number of employees on the company's books, not to mention its profits, hardly justified the expense of a Managing Director, the scope of its operations could hardly be called limited. Indeed, the company's business became increasingly complex as it diversified into many areas unrelated to the wool trade. This was recognised in 1976 when the Chairman remarked on the difficulty of obtaining information on subsidiaries and associated companies. J. A. Jowitt proposed, seconded by Richard Heaton, that a Book-Keeper and an Assistant Secretary be recruited to handle the extra work.[837]

DIVERSIFICATION

The first of the new diversification proposals were slow to emerge and were far from radical, in that they were still within the area of wool or other textiles. The first was the possible investment in Monaco Manufacturing (Household Textiles) Ltd, a producer of fitted covers, which foundered on what was considered an unreasonably high value placed on the company by the proprietor.[838] Monaco approached the company again in late 1971 seeking finance but were told that Jowitts would only consider investment after restructuring had taken place.[839] It went into liquidation in 1973.[840] The second was the possibility of entering the double jersey knitting field, an area which had been booming for some years. The Chairman believed that the use of knitted fabrics for men's outerwear would be a growth area in which the company should concentrate.[841] Knitted garments, which suited the seventies' less formal dress code, could, in contrast to tailored garments, be manufactured in a highly automated way with almost no wastage.

Peter Adams Agencies for Textiles Ltd

In 1971, F. T. B. Jowitt was approached by Peter Adams, a cloth-manufacturer's agent, with a proposal for Peter Adams Agencies for Textiles Ltd.[842] The company financed an initial trip to the Middle East which generated sales of £80,000. For the year to 31 October 1973, Peter Adams Agencies had a turnover in excess of £1 million, although it reported a loss due to the delay between sales and receiving commission payments.[843] The Agencies grew to include new territories in Scandinavia, South Korea and Argentina. However, after heavy trading losses at Hollings Mill, the company decided to sell its interest to Vitale Barberis for £8,250 in 1978, netting a profit of £7,500.[844] The sales contract stipulated that if Barberis sold Pater Adams Agencies on Jowitts would receive a percentage of the sale price, which brought the company a further profit in due course.

Everfair Ltd

They were also approached by Mr E. Stern who had been employed by Illingworth Morris for nine years as Director of Scott Woodhead Ltd, cloth manufacturers without looms – that is, they commissioned cloth to be made to their specifications. Mr Stern had been very successful in his previous employment and the company wasted no time in forming a joint venture with him, Everfair Ltd, in March 1972, giving Barclays Bank an unlimited guarantee to secure the new company's account in April. Everfair had already obtained orders from the British Overseas Airways Corporation (BOAC) by March 1972,[845] and its first three months of

Fig. 45 In 1975 Robt. Jowitt & Sons celebrated what was considered to be the
company's bicentenary with a dinner. The sorting room proved an ideal ballroom.
From left to right: David Foster, Susan Foster, F. T. B. Jowitt, Juliet Jowitt,
Douglas Barker and Mary Barker.

trading produced a profit before overheads of about £4,000. This was
deemed very satisfactory, especially since the company had been faced
with a steep rise in yarn prices.[846] The company continued growing, with
a profit for the 1973 financial year of about £17,000 before Stern's bonus of £2,000 and Jowitts' management charge of £6,000 and another
good profit in 1974.[847] After a dip in both turnover and profits in 1975 and
difficult trading conditions in the early part of 1976, Everfair turned in
improved results for the 1976 financial year.[848] The company continued
to trade well in 1977, but Mr Stern suffered from a stroke and there was
doubt as to whether he would ever return to work full-time. Despite this,
the company made a profit of about £24,000 in the first six months of
the 1978 financial year. Worryingly, Jowitts was experiencing difficulty in
finding an assistant to help Mr Stern. The company finished 1978 with a
profit of £29,000, slightly ahead of forecast.[849] The profit for the follow-

ing year, at £21,000, although reduced, was considered reasonable in difficult trading conditions. However, the company was still reliant on one big customer, British Airways (a merger of the former BOAC and BEA) and F. T. B. Jowitt expressed doubt as to whether the company would be able to expand its customer base.[850]

Mr Stern died in June 1982 and his place was taken by D. H. Foster's brother John. Jowitts made an offer for Stern's share of the company which was not well received by his family.[851] Facing a large loss on its wool business in 1982, Jowitts resolved that Everfair should pay a large dividend, effectively transferring all its reserves to the mother company.[852] Everfair continued to trade well over the next few years and managed to diversify, especially in the supply of cotton check linings for (amongst others) Barbour waxed jackets, nominally manufactured in Hong Kong but in reality sourced from mainland China. Even so, the loss of the British Airways account in 1986 hit it hard, with the anticipated turnover falling from £300,000 to £180,000.[853] In 1986 the Jowitt Board decided that Everfair was too small to make sense as a separate company and it was absorbed into Robt. Jowitt & Sons Ltd, producing a useful saving in audit fees.[854]

Property

As the company's wool operations contracted, office space at 153 Sunbridge Road became available which was rented out to a number of organisations, including the Department of Social Security. In June 1972, Peter Bell suggested buying the adjacent property, 149/151 Sunbridge Road. He believed that, after suitable conversion, it could be let very profitably. The Directors sanctioned spending up to £30,000 on this building and it was eventually bought in March 1973 for £27,500, with an additional £80,000 being allowed for refurbishment.[855] Even at the relatively low bank rate of 6% prevailing in 1972,[856] returns would have to be impressive just to pay the interest bill. The property was found to have structural problems, notably corroded rebars, and had not been let by mid-1974.[857] It was sold in 1983, the company receiving £13,073 less than it had paid for it. In reality, though, the property had proved a very good investment. It had land to the rear which provided valuable car parking space for the tenants of 153 Sunbridge Road. In addition, 151 and 153 Sunbridge Road stood on either side of Greenacre Place, a no-through road, which the company was able to close and use for additional parking spaces.

A Miniature Venture Capital Company or Mini-Conglomerate?

Despite Barclays' reservations, Robt. Jowitt & Sons Ltd was transforming itself from a wool company to a miniature venture capital company or mini-comglomerate, with investments in a diverse range of companies and other assets. In December 1972 it made a loan of £5,000 to Heppenstall Commodities Ltd, decided to buy up to £30,000-worth of shares in Rutland Estates and Trading Co. Ltd (see p. 226), a caravan site company, agreeing to lend that company up to £20,000, and bought a cottage near Fountains Abbey. At the same time it sold shares in a Leeds company, Farnell Electronics Ltd at a small profit and bought shares in Courtaulds Ltd and John E. Dallas & Sons Ltd, a musical instrument manufacturer. In March 1973, the company bought shares in Cavendish Land Co. Ltd. All these, as well as the company's commitments to Everfair, Peter Adams and the purchase and conversion of 149/151 Sunbridge Road, were largely financed by bank borrowings,[858] and the company had to use shares in Courtaulds Ltd and Sirdar Ltd, a supplier of knitting yarns, as security for the borrowing.[859]

Between 1 November 1973 and 26 June 1974, the company bought Kruger Rands, bought and sold gold sovereigns at a handsome profit, bought and sold shares in London Bridge Securities at a substantial loss, and bought shares in Southern Pacific Properties Ltd, Debenhams Ltd, London & Overseas Freighters Ltd and Northern Capital Ltd.[860] Between June and October 1974, the company bought twenty-five Kruger Rands and shares in Consolidated Goldfields and rebought shares in Sirdar Ltd at 18p per share. These were sold in the following years at a significant profit. On 30 October 1974, it sold shares in Valor Co. Ltd at a loss of £982.34 on an initial investment of £1309.30 and shares in Courtaulds at a loss of £2,257.40 on an initial investment of £3,600.10, losses of a remarkable 75% and 63% respectively.[861] In 1975 it lent £10,000 to D. G. Barker & Co., an agency for Caravelle freezers owned by the Company Secretary's son. It also subscribed to £2,000-worth of shares in Bradford Community Radio,[862] which proved to be a profitable investment.

Rutland Estates and Trading Co. Ltd

Rutland Estates grew out of a relationship that Peter Bell and F. T. B Jowitt had with Alan Hird FCA and Roger Suddards CBE, DL. Roger Suddards was a great friend of Mr Jowitt's, Pro-Chancellor of Bradford University, a Governor of Bradford School, a scion of a prominent Bradford legal family and finally a partner of the solicitors Last Suddards and Company. Both Alan Hird and Roger Suddards had taken offices near to the Jowitt offices in Sunbridge Road so it was natural they and the Jowitt directors saw a lot of each other. All four shared a passionate interest in the wellbeing of the Bradford community and especially in the commercial aspects. In particular, they were keen to foster a relationship between the University and the business community and to seek local investment opportunities. If they were considered suitable, they could be financed and developed locally before being brought to the market. The vehicle created to implement this plan was Northern Capital. Sadly, despite many discussions, the University failed to come up with any suitable ideas and what seemed like a promising venture did not take off.

Northern Capital was sold to a Mr Benjamin and little more was heard of it. Nonetheless, the association between the four men led to the creation of one of Jowitts' less likely investments. Alan Hird had two clients, Terry and Phil Duggan, Bradford (or, more particularly, Baildon) builders who had acquired a caravan park called Hawkesworth and the lease of a second, council-owned park called Baildon Moor. They wished to expand and Roger Suddards had learnt that the well-established and well-run Cringles caravan park was coming to market. Robt. Jowitt & Sons was able to buy Cringles and use it as a bargaining chip in a proposed joint venture with the Duggan brothers. A deal was struck and Rutland Estates was created with Roger Suddards as Chairman. The Duggan brothers, F. T. B. Jowitt, Peter Bell and Alan Hird were all appointed to the Board. After initially appointing a manager who proved ill-suited to the task, the Board appointed Anthony Chisenhale-Marsh, a friend of both F. T. B. Jowitt and Charles Leveson-Gower and, as the former manager of a division of Charringtons brewery, a man of considerable business experience. The intention was to expand the enterprise while bringing management expertise and strong branding to it. In the event, the company found it difficult to operate in the unfamiliar milieu of the Romany community which, at the time, dominated the industry. Despite Roger Suddards's expertise in planning law, the company also experienced persistent problems with the planning authorities. Although the caravan park company made no major operating losses, neither did it make substantial profits. In May 1978, Mr Jowitt proposed to liquidate Rutland Estates and hive off Cringles into a separate company in which Jowitts would take a 50% stake.[863] The hiving-off went ahead, although liquidation of Rutland's assets was still not complete in September 1979.[864] With the liquidation still far from complete in May 1980, Jowitts increased its holding to 50%, certain that the underlying assets more than justified the cost .[865] The liquidation was still not complete in February 1983, but the Directors maintained their belief in the strength of the underlying assets.[866]

While Cringles continued to trade at a reasonable level, Mr Leveson-Gower questioned in January 1983 whether it was a useful investment to hold and it was agreed that he should approach the owner of the other half share, Mr Chisenhale-Marsh to see if he would be interested in buying Jowitts out. However, Mr Chisenhale-Marsh had received a valuation of £82,500, which he considered unrealistically high as alterations were needed to the sewerage system.[867] At this sort of price, there was little incentive to sell, but the company obtained a valuation of £170,000 from a specialist and Mr Chisenhale-Marsh indicated that he would wish to sell if such a price was achievable. The company prepared to auction the site in Autumn 1984, but before this could happen an offer was received for £165,000, which was accepted. Further assets of Cringles Caravan Park Ltd were estimated to be worth £33,200.[868] Jowitts made a very substantial capital profit on these asset disposals.

Both Roger Suddards and Alan Hird continued in close association with Jowitts. An amiable and universally respected character within the community, Roger Suddards also showed a swift, practical response to a crisis. When a dreadful fire at Bradford City AFC's Valley Parade stadium killed 56 people and injured at least 265, on 11 May 1985, Suddards immediately swung into action. Working with Gerald Hodges and Keith Marsden, he raised £4.25 million from the community and distributed it to the injured and bereaved with exemplary speed. Suddards was subsequently invited to 10 Downing Street to discuss his ideas on emergency planning with Margaret Thatcher. Sadly his death in 1995 at the age of sixty-five deprived the community of a great leader and benefactor.

It was felt that the Board of Jowitts lacked accountancy expertise; although Norman Matthews was a qualified company secretary, he was not a chartered accountant. Accordingly Alan Hird was asked to join the board in 1989 to bring his special and alert knowledge to its deliberations which he did. It was at his suggestion that the company attempted to eliminate the preference share capital which, it was felt, only cluttered the balance sheet. However, this was not universally welcomed by the shareholders and was only partially achieved, thus making matters rather worse. He died in 1992 at the age of fifty-eight.

Solar Water Heaters Ltd

Far ahead of their time in terms of environmental awareness, the Directors sanctioned a loan of £2,500, secured against existing orders, to Solar Water Heaters Ltd (SWH), a company founded by John Cook, a bright young engineering graduate, in 1975. Unfortunately, the original panels which the company used were found to be difficult to assemble and it was considered expedient to source a container-load of better ones from Fafco in the United States for $30,000.[869] The panels were primarily designed for heating swimming pools but could also be adapted to domestic heating. In the cloudy British climate, efficient domestic water heating was difficult to achieve in the winter months, whereas swimming pools, used almost exclusively during the summer months, were a more practical proposition.

While a great deal of interest was shown in the technology, few orders were received and in September 1975 it was agreed to keep SWH financed up to May 1976. It was also agreed to appoint a salesman.[870] Sales increased but profit margins on the panels, which were purchased through Discus, an intermediary, were very poor[871] and the Board agreed in September 1976 that F. T. B. Jowitt and John Cook of Solar Water Heaters should approach Fafco in October about the possibility of producing the panels in Europe. Solar Water Heaters had accumulated losses of about £44,000 by the end of the 1976 financial year.

In February 1977 it was reported that Fafco had severed its relationship with Discus and reduced the unit cost to $55. At this price, the break-even point was estimated to be about 1,700 units per annum, which was thought to be achievable. In fact, one customer, CTT, had indicated that it would want 800 units in May. Solar Water Heaters had hoped to expand into Spain but, with Fafco having its own plans for a European operation, it appeared that SWH would have to restrict itself to the UK and Ireland.[872] There was some talk of Fafco setting up a production facility in the UK, in which case Jowitts might participate by investing 40% of the £100,000 start-up cost.[873]

Unfortunately, Solar Water Heaters' sales continued to disappoint. For the ten months to 31 August 1979, the company showed a pre-tax loss of £8,203. The general view of the Jowitt Board was that the SWH could not continue to trade in its present form. However, F. T. B. Jowitt pointed out the difficulty of winding up the company and drew the Board's attention to SWH's stock, valued at £31,839. He also felt that Mr Cook's wealth of knowledge built up with SWH could be valuable in the area of energy conservation. It was agreed to keep the company trading for a further six months in order to sell off the stock at reasonable prices. It was feared, however, that the end result would still be a write-off of the investment.[874] The figures for the financial year to 31 October 1979 showed a loss of £8,600 before interest payments.

While the stock was being sold off, Solar Water Heaters investigated the use of solar panels to pre-heat water for domestic central systems. It also started a sideline in installing cellulose loft insulation.[875] A surprise surge in demand for the solar panels in early 1980, just as SWH had sold its remaining stock, encouraged the company to order another container-load[876] and by June it was predicted that the company would return a profit, since it had raised its prices and was benefitting from a favourable exchange rate against the dollar.[877] Solar Water Heaters returned its first profit in the 1980 financial year, although far less than the £20,000 which had been predicted earlier in the year.[878] Loft insulation was briefly profitable but the work disappeared as soon as government grants ran out,[879] reviving just as rapidly when the government made new grants available in December 1980.[880]

The results for the half year to 30 April 1981 showed a loss of £16,000 and it now seemed clear that the profit in the second (usually more profitable) half of the year would be lower than predicted. Computerisation of the roof insulation side showed that the profit margin on this was lower than expected and would only contribute about £6,000 for the year, whereas the solar panels were expected to make a loss of £15,000. The Directors concluded that there was 'no real market in this country' for solar panels and that SWH must trade out of this market, either through retail sales or by selling the remaining stock to another distributor.[881] The plan to use the panels for domestic water heating never came to fruition and a separate company intended to handle this area, Solar Domestic Products Ltd, never traded.[882] By January 1983, Solar Water Heaters was reduced to an insulation company only and was expected to make a profit of £10,000 after bearing charges of £12,500 from Jowitts. The staff had been reduced to a minimum, with only two salesmen operating, and its continued trading was described as 'precarious'. However, the Company Secretary said that it could be closed down quickly without any loss to Jowitts.[883] The following month, it was announced that SWH had a deficiency of nett asssets. The Board decided that Jowitts could no longer provide financial assistance, knowing that this refusal would cause SWH's accounts to be qualified. Quite unexpectedly, Solar Water Heaters received a contract to insulate 600 homes in Leeds and also sold a number of solar panels which had been written-off at the end of the previous year, with the result that it contributed £43,000 to Jowitts' half-year profits.[884] After that, however, orders for insulation disappeared and the staff were made redundant. There were still large stocks of solar panels. Jowitts intended to sell off the remaining panels, subcontract any remaining insulation work and wind up the company by the end of the 1983 financial year, with all outside creditors being paid. The capital losses on loans made to the subsidiary would be used to offset profits made on the liquidation of another subsidiary.[885] In reality, the winding-up took until late 1984.[886]

Bottle Recovery Ltd

Another 'green' investment was made in early 1976, when the company agreed to provide up to £50,000 finance for Bottle Recovery Ltd, a company (then in receivership) which collected and recycled wine bottles.[887] It was a service very much of its time, sandwiched as it was between the period when deposits were commonly charged on bottles and the present era of mass recycling. Unfortunately, it was a much more complex and laborious process than, say, a brewery collecting its bottles for re-use, or the process of recycling scrap glass. Bottle Recovery's aim was to collect bottles in bulk from business premises and then clean and sort them for re-use by many companies. The Jowitt Board noted in September 1976 that there were problems with the cleaning, sorting and management.[888] A particular problem was encountered when some bottling plants started employing latex-based adhesive which were extremely difficult to remove. Nonetheless, in the year to 31 October 1977 it made a profit of about £5,000.[889] Although Bottle Recovery was never a great success, the financing from Jowitts was repaid in full by the proprietor, Guy Norton.

Sunbridge Electrical Wholesalers Ltd

In March 1976, the company received a proposal from Sir Charles Legard Bt to invest in a company selling electrical control panels to industry. The panel would be supplied by Télémécanique Eléctrique (G. B.) Ltd. Subject to satisfactory assurances, the company agreed to invest up to £15,000.[890] Sunbridge Electrical Wholesalers Ltd was duly set up with a staff consisting of a manager and private secretary and was granted the territory of West Yorkshire excluding Leeds by Télémécanique.[891] The profit in the first fifteen months of trading, to 31 October 1977, was hardly spectacular at £7,000 but was quite acceptable for a start-up.[892] The company grew rapidly, especially after Leeds was added to its territory in 1977. In 1979 it added the distributorship of Watford Electrical Products Ltd and returned a profit of £31,000 before bonuses, Directors' fees and management fees.[893] The first six months of the year to October 1980 were very good, with the company running at £30,000 ahead of forecast, but difficult trading conditions and a bad debt of £6,000 caused the company to finish the year well below its anticipated profit.[894] The company recovered in 1981 but started to experience difficulties again in 1982. The profits for the first nine months of the financial year were only £7,000 against a forecast of £12,000 and Legard was expressing his concern over the level of stocks and debtor balances which seemed to consume the profits. He expressed his wish to disengage himself from the company, which was taken by Jowitts as a tactic to encourage the company to tackle these problems.[895] However, Legard did reduce his activity in the company and, as he was the expert in electrical control systems, this was seen by Jowitts to be a factor inhibiting the company's growth.[896] Perhaps as a result, the company finished the 1983 tax year just below forecast.[897] Despite occasional downturns and bad debts, Sunbridge Electrical continued to perform well, so well, indeed that in 1986 the Jowitt Directors decided to look for other electrical acquisitions, although they were concerned that such a move might not meet with Legard's approval.[898] This enthusiasm for Sunbridge Electrical seemed to be vindicated when Jowitts' 45% share of its 1986 profits amounted to £31,000 against a forecast of £22,000.[899] Its contribution in the following year was estimated at over £65,000, but fearing its over-reliance on Téléméchanique, an effort was made to merge with Barron Control Gear and also with Sir Charles Legard's own Electromatic Engineering Ltd. It was also decided to open a branch in Chester to cover Wales, Ellesmere Port and the Wirral, but in the event a branch was opened in Deeside instead, for which area the company obtained an agency agreement with GEC.[900] The Deeside branch got off to a slow start and, rather worryingly, was achieving margins of 14.3% compared with 17–18% in Bradford.[901]

The annual turnover had grown to £2.41 million for the 1988 financial year but margins had slipped resulting in a profit of about £110,000. The Jowitt Directors expressed concern that Fred Etherington, the Managing Director of Sunbridge Electrical (on secondment from Electromatic Engineering), was only spending one day a week at the company which seemed disproportionate to the size and expected growth of the company.[902] To encourage his greater attendance, or at least avoid paying for his absence, Jowitts agreed a temporary arrangement with Sir Charles Legard by which Etherington's time might be charged on an hourly basis.[903] In October 1990 Jowitts tabled an offer to buy out Legard's share. Unfortunately, Robt. Jowitt & Sons' finances did not allow it to make a full cash offer and Legard was not attracted by the deferred payment terms. He then made a counter-offer to buy the Jowitt shareholding. While the initial offer was considered too low, Jowitts eventually agreed to sell at the end of May 1991 for £200,825, with the company also receiving a dividend of £93,333.[904] It was the end of Robt. Jowitt & Sons' most inspired and most profitable diversification, although the company parted with it rather cheaply at little more than twice earnings.

Preston Gardner (Commodities) Ltd

In 1981 Mr F. T. B. Jowitt suggested to the Board of Robt. Jowitt & Sons that the company might be interested in investing in Preston Gardner. He explained that this commodities broker had been founded by M. C.

Brackenbury & Co.,[905] Michael Preston Gardner and himself in 1976. It had originally been a low-key affair managed by Mr Gardner and himself in their spare time. However, it had grown and Mr Gardner had left Bradford to work for Brackenburys in London, leaving F. T. B. Jowitt with a heavy demand on his time. The existing proprietors were reluctant to give up this apparently profitable enterprise but, although it was clear that a manager should be appointed, it would be difficult to attract a suitably-qualified person to such a small company. It was also thought that some money should be spent on promotion. M. C. Brackenbury & Co. had indicated that it was willing to buy 50% of the share capital and provide a loan of up to £20,000 if Robt. Jowitt & Sons were willing to make a similar commitment. Mr Jowitt emphasised his own interest in Preston Gardner and stressed that, if the company were to invest, it should not pay more than the nett asset value.[906] He further warned that he was quite unable to predict future profits and also wondered if there were not a contradiction between the Board's wish to invest in Preston Gardner and its reluctance to be involved in speculative commodities trading. This is slightly misleading as the company had been involved in some fairly speculative trading of commodities futures. In the previous year, Mr F. T. B. Jowitt had told the Board that, in order to recoup losses of £21,000 on the Sydney Futures Market, the company had bought sugar futures, adding that

> it was not the intention of the executive directors that the company should trade on the commodities futures market on a regular basis.[907]

Unswayed by the Chairman's warning, the Board decided to make the investment[908] and F. T. B. Jowitt and G. N. Inglis, the Company Secretary, were nominated as the company's Directors on the Preston Gardner Board.[909]

Preston Gardner made a loss in 1983 and the Jowitt Board concluded in January 1984 that there was little prospect of it breaking even from its existing activities. A new fund had been devised to allow individuals to invest relatively small sums in gold and Jowitts decided that if there was no prospect of attracting at least £100,000 of investment by April, it would advise M. C. Brackenbury & Co. of its wish to withdraw from the venture.[910] Despite failing to reach this target and incurring liabilities due to 'mistaken handling of clients' funds,' Preston Gardner enjoyed a good month in May and Jowitts decided to allow it to continue trading until the September Board meeting.[911] Although it was reported in September that Preston Gardner had made trading losses of £27,000 in the ten months to 31 August through trading losses and bad debts, and that it had lost most of its clients, it was decided to keep it trading in the hope of clawing back some of the losses. The staff had been laid off and F. T. B. Jowitt was again managing the company so that overheads were at a minimum.[912] Remarkably, the company then started to make a small positive contribution on the basis of its few remaining clients.[913] In September 1986 it was forecast to make a profit of £15,000 on commissions from just two or three clients.[914] At this stage, the company sold its stake and control passed to M. C. Brackenbury & Co. with Mike Gardner running it from their office in London.

Computers and Computing

In the late fifties, the most advanced accounting equipment available to most companies was a mechanical adding machine, such as the Sum-Lock purchased by Jowitts in 1959 for £160.[915] By 1967, the company had acquired an Olivetti desktop computer for £1,750.[916] However, the investment in Sunbridge Electricals in 1976 not only heightened the Directors' awareness of the 'microprocessor revolution' (as the Company Secretary, G. N. Inglis, called it at the time), it and the other acquisitions also increased the demands for a substantial computer system. After investigation, it was decided to lease equipment produced by the German Company Nixdorf. It would cost about £9,000 per annum and would offer savings and benefits to Jowitts of approximately £5,000, plus about £2,000 to Sunbridge Electricals. To sweeten the deal, Nixdorf offered to make Jowitts a demonstration site for its computers with a 5% commission payable on any resulting sales. A guaranteed commission of £5,000 would be paid for the first two years.[917] There is no indication that any money was made from this arrangement.

In 1990, Alan Hird suggested that the company might be able to profit from the growing popularity of personal computers by lending £20,000 to Access Publications, publisher of the proposed *Access* popular computing magazine.

It was hoped that the company could quickly sell off its stake in Access Publications at a substantial profit. Unfortunately, the very first issue in October that year went out with a faulty tape cassette mounted on the cover. Although the second issue was more successful, the company never recovered from the initial failure and Mr F. T. B. Jowitt reported in January that a creditors' meeting had been called and that there was little hope of recovering any of the loan.[918]

Other Investments

Robt. Jowitt & Sons invested in a perplexing variety of businesses, including the strange excursion into kangaroo fur coats sold under the 'Outback' label. They were designed by a man called Bond and the subsidiary's dialing code was 007 but despite this and a flurry of enthusiastic press coverage, interest soon faded. There were so many bright ideas which, for whatever reason, failed to transform the company's fortunes that it would be impossible to cover them all. Closer to home, a great deal of time and effort was expended on the Sunbridge Road properties, with the company renting out more and more of its office space out as its own needs decreased.

OVERSEAS BRANCHES

By the late seventies, the company's branches in Australia and South Africa had ceased all activities other than the rental or sale of its properties and investment of the proceeds.[919] The last of the Australian properties, in Fremantle, which was being leased to Plimex (Pty) Ltd, was sold to Victoria Investments Ltd in Spring 1981 for $A210,000.[920] The proceeds were invested through T. C. Coombs, largely in Australian shares, with the intention of generating growth rather then income.[921] The directors were unamused to discover in June 1984 that the investments had lost £10,000 in value, the loss growing to £23,000 by September. However, this was caused by the suspension of shares in Allied Resources by the Australian Stock Exchange. When the suspension was lifted, the investments recovered their value.[922] The company sold off the investments in Autumn 1986.[923]

The South African situation was rather more complicated. The company had planned to sell its South African subsidiary, Sunbridge Properties (Pty) Ltd, to the existing tenant of its Durban property in 1978 but the South African Reserve Bank[924] had indicated that exchange controls would oblige the company to reinvest the proceeds in low-yielding South African government bonds which would allow the money to be repatriated on maturity. The company then decided to sell the property and invest the proceeds in some alternative form of security,[925] but had great difficulty in selling. It was advised in 1981 by an estate agent that the poor state of the building and the poor return on the lease, which still had five years to run, made it impossible to sell at a reasonable price. He advised that the company would receive a far better return if it renovated the property and then waited until there were only two or three years left on the lease.[926] The sitting tenant, Mr Xhaust, made an offer of R175,000 (against a book value of R145,000) for the property in late 1982, which the Chairman considered inadequate. However, the Secretary, C. N. Inglis, pointed out that a sharp fall in Sterling made this price much more attractive. Following the De Cock Commission Report in 1983, the South African Government lifted exchange controls. F. T. B. Jowitt, fearing (rightly, as it transpired) that this would not last, sold the property in March 1983 for about R190,000. With a very favourable exchange rate of about R1.65 = £1 (the current rate is about R11 = £1), he repatriated the money at a Sterling value well in excess of the book value. Sunbridge Properties (Pty) Ltd was then liquidated.[927]

THE WOOL BUSINESS

Although the Chairman had emphasised his belief that woolcombing in the UK was not an area which could be relied upon to provide shareholders with an adequate return, there were good reasons for the company to continue working in some sort of wool-related activity. Not only did it have a wealth of experience, equipment, warehouses and an excellent reputation in the wool trade, it was also constrained by what Barclays Bank would allow it to do, the bank having made it clear that the company's borrowing facilities were 'not available for diversification without prior discussion'. There were also significant tax losses amounting to over £¼ million which could only be set against wool trade profits.

The End of Woolcombing at Hollings Mill

Since the closure of the Meanwood fellmongery in 1967, the company had been left with the wool merchanting business, mohair, topmaking/woolcombing and the experimental shrink-proofing process. All these areas looked vulnerable, but none more so than the combing plant at Hollings Mill. In July 1970, F. T. B. Jowitt told the board:

> ... the stock of finished tops was increasing as a result of diminishing delivery orders and that, although the order book was reasonably long, if the rate of off-take did not improve it would be necessary for the mill to stand for the whole of August. He explained further that a menacing situation was becoming apparent where the rate of sales was dropping below the output capacity of Hollings Mill. Up to recently, the Company had always been able to sell tops well in excess of its combing capacity and thus has maintained efficient machinery utilization in the face of fluctuating demand. It appeared that this happy situation no longer obtained.[928]

In March 1971, the Managing Director reported that the mill was only running three or four days per week and would only break even at this rate, compared with a profit of about £70,000 per annum when running at full capacity. Moreover,

> It appeared that the reduced world demand for fine wools was to be a permanent feature of the textile scene. If this was so, the stocks of these wools in Bradford and in the course of accumulation [in support of reserve prices] by the A. W. C. [Australian Wool Corporation] was a source of concern in Britain, and particularly to the Company who historically had concentrated on wools of 64's quality.

> With trading conditions likely to continue to be bad for the remainder of the year, Mr P. J. M. Bell could not see the Mill going back to full production. In view of this, it was essential to scrutinize overheads again and also consider further redundancies in order to bring overheads more in line with the Company's reduced turnover.[929]

Accordingly, the whole night-shift was made redundant at the end of April 1971.[930] Robt. Jowitt & Sons Ltd was not the only woolcombing company to be suffering and in May the Managing Director informed the Board that Airedale Combing Co., the largest single combing mill in the UK, was to close as the plant had been bought by the Woolcombers' Mutual Association. At this point, it was decided that the Robt. Jowitt & Sons should offer its own combing plant to the WMA for £110,000.[931] When the WMA offered £80,000, the Board considered that it had three possible choices: it could sell to the WMA; it could keep the combing plant without modification; or it could keep the plant and make modifications to create a mixed plant of Noble and Schlumberger combs which, the Managing Director estimated would cost £30,000. The Board also contemplated the liquidation of the company after a sale of the combing plant to the WMA but this was rejected unanimously, not least because it would mean wasting the company's tax losses. The directors concluded that sale of the plant was, indeed, the best option and authorised the Managing Director to negotiate with the WMA 'on the best terms possible but not less than £80,000'. The additional clause seeking 50% of the realised selling price of the plant was not accepted by the WMA. The Chairman drew the Directors' attention to the number of long-service employees who were likely to be affected by this decision and deep regret was expressed by the whole Board. The Managing Director was instructed to accord redundant staff 'the same generous compensation as had been paid to previous cases of staff redundancy'.[932] The sale was concluded in December 1971 for £95,000 to be paid in three equal installments in October 1971, March 1972 and September 1972 plus interest at 6¾%, a surplus of £46,700 over the book value, although £33,600 was absorbed by redundancy payments. These payments included £7,641 made to J. B. and Geoffrey Harland who had resigned with effect from 30 April 1971.[933] It was the end of the Harland family's long association with Robt. Jowitt & Sons. Geoffrey Harland, deeply resenting his treatment, bought the Buck Inn at Malham. Some two years later, Peter Bell was lunching with the head of the Australian Department of Agriculture in a local restaurant when Geoffrey Harland appeared. Peter feared an acrimonious confrontation but was greeted warmly and thanked for giving Jowitts' former Topmaker the opportunity to build a wonderful new life. Multiple reports confirmed that Geoffrey had made a great success of the Buck Inn. Sadly, J. B. Harland died of cancer in 1977.

While the company was moving out of combing, it was retaining its topmaking activities. That is, it would buy and sort wool which would then be sent for commission combing and sell the resulting tops. It was said that, with the reduced overheads, topmaking should remain economic if the company could retain its existing customers and achieve a turnover of £¾ million per annum. It would employ about £250,000 capital.[934] The company also maintained its wool top shrink treatment plant and the sale excluded machinery for this purpose. Investment was made in a Petrie McNaught backwash modified for the shrink treatment process, together with the necessary chemical storage and metering equipment. Mr Bernard Hankin, the company's electrician, was appointed manager of the new department.

Out of combing themselves, Jowitts encouraged the creation of Yorkshire Combers Ltd in 1973 by Lord Barnby, Francis Dawson of H. Dawson & Sons and others.[935] Although ably led by Stuart Twitchell as Managing Director, this venture faced the same difficulties as the rest of the industry and was relatively short-lived.

The End of Trading in Wool and Tops

Even topmaking and general wool trading became increasingly unviable in the late 1970s. Sirdar PLC, a well-known producer of hand-knitting yarns and one of the company's most important customers, had failed to take up contracts. F. T. B. Jowitt had tried to interest them in fixed-price contracts but they baulked at the very modest premium required by Jowitts to insure themselves against adverse price movements. A slump in the worsted trade in 1978 was described as 'the most serious in living memory,'[936] and in March 1980 C. M. Leveson-Gower went so far as to question

> ... the wisdom of continuing to trade in wool and tops, since it appeared that profits painfully accumulated in other sections of the business were being lost in this, the company's original activity. He proposed, with the support of Mr Inglis [who had been appointed Company Secretary on 1 June 1977 and Director on 26 September 1977] and Mr Foster, that the company should dispose of its stocks of wool as soon as possible and withdraw from activity both in physical wool and in futures. After some discussion, the Board resolved to adopt this course of action.[937]

Despite a very sluggish market, much of the stock had been disposed of by September[938] and, except in the specialist areas mentioned below, the company's long association with wool came to a close.

Serico-Lana

As well as the shrink treatment of wool tops (see below), the company invested some time and money on another IWS-developed process called Serico-Lana which turned wool into a smooth, glossy, silk-like fibre. Although it was well received by spinners and retailers at the British Yarn Show in 1991,[939] it was not ultimately very successful.

Superwash Wool Top Shrink Treatment

While the company was moving out of combing, it continued to develop wool top shrink-treatment technology, which, at the end of 1971, had been licensed by the IWS to Jowitts and Woolcombers Ltd. The resultant tops were fully machine-washable and expected to be particularly suitable for the jersey machine-knitting yarns and hand-knitting wools. Indeed, the Chairman explained later that the main reason that the company committed to the process in the first place was to avoid losing its major customer, the hand-knitting yarn company Sirdar.[940] After the sale of its combing plant to the WMA, Jowitts had retained a certain amount of gilling equipment to re-align the fibres after shrink treatment.[941] The shrink-treated product was launched in March 1972 under its International Wool Secretariat 'Superwash' licence, a term also applied to wool treated by the competing Dylan process.[942] Although the product was well received, the Managing Director was already reporting slackening interest in Superwash by June[943] and, after an interest charge on the capital employed and an allocation of £10,500 for general overheads, the shrink-treating lost £38,000 for the year to 31 October 1972. Although Sirdar was buying all its Superwash treated wool tops from Jowitts, it was ordering far less than anticipated. The Chairman reported that the IWS had launched an advertising campaign for Superwash aimed at promoting sales to hand-knitting companies.[944]

It appeared that the high price of wool was encouraging the substitution of synthetic fibres and also that manufacturers were still choosing to apply shrink treatment at the fabric stage rather than to tops prior to spinning. As a result, by June 1973, the mill was still not in profit and the company was talking of writing off the whole cost of the shrink-treatment plant during the financial year.[945] With the plant still running but sustained by commission gill blending of wool tops with synthetics, the plant was reported to be very busy by June 1974, although it was uncertain how long this would last.[946] Happily, in December the plant was working at full capacity and was fully booked until April 1975 and it was still at full capacity in September.[947] Such was the sudden success of Superwash, with Patons and Baldwins, another major hand-knitting yarn company, forecasting a requirement of up to 14,000 kilos per week, that by December 1975 the Board was contemplating a second line costing £70,000.[948] This plant was ordered but only delivered at the end of the 1976 financial year, later than expected, leading to lower Mill profits than had been anticipated. The company was also experiencing difficulty gilling the treated tops, but a work-study was underway with the hope of increasing the efficiency of the whole plant. Despite these difficulties, the plant was reported in December 1976 to be fully booked until April 1977.[949] Buoyed by the high demand, the company investigated the use of a machine designed by Kroy Unshrinkable Wools Ltd of Toronto which was said to reduce start-up time and increase processing speed, thus reducing total costs despite charging a royalty of 4p per kilo. It was hoped that a machine might be loaned for a few months to allow for assessment.[950] Despite falling demand in the early part of 1978, the Chairman and W. J. B. Jowitt visited Kroy where they negotiated a deal to acquire a Kroy chlorinator. As it was not in commercial use anywhere, the company took it for a three-month trial, able sell it back to Kroy at the purchase price should it not perform to specification.[951] While the plant was once again fully-booked by September 1978, the Kroy machine was not working to specification and the manufacturer was hoping to remedy this with modifications,[952] although there had still been no progress by December.[953] A replacement Kroy machine was finally delivered in 1981 and by July it was showing useful economies at high volumes.[954]

The company had enjoyed an exclusivity agreement for the UK with Kroy which ran until 1 September 1985. Since this process gave Jowitts a strong competitive advantage, it might be assumed that they were in a weak bargaining position. David Foster visited Kroy in Canada and, remarkably, managed to negotiate a new three-year agreement on more favourable terms.[955] This did not stop the Chairman and Stephen Lambert, Bernard Hankins's successor as plant manager, from visiting Switzerland to investigate the rival Fleissner chlorination process. Their report seemed to show that the results were as good as but no better than those from the Kroy process, and there was some concern that it might cause some fibre damage. The Swiss machinery was built to the highest standard, while Kroy's was said not to be very well engineered. However, the Board realised that they had worked hard to achieve acceptance of the Kroy product and it would take effort to convince customers of the value of a new process. While not placing an order, the Board decided to sign a letter of intent.[956]

Mill activity increased substantially in 1987 and the company received a large order from the German company Schachenmayr Mann, encouraging it to contemplate purchasing another building at a cost of £350,000.[957] However, as was so often the case in the wool trade, things had altered radically by December when the company lost the German contract, due to a rise in Sterling, the collapse of the German hand-knitting sector and disgreements with Schachenmayr Mann over shrink-proofing standards,[958] in particular the pH of the treated tops. Since the plant manager, Stephen Lambert, a former dyeing manager and experienced chemist, was perfectly capable of neutralising the alkalinity of the chlorine-based process the Jowitt directors suspected that the complaint was dubious, especially as Schachenmayr was known to be in financial difficulties. In the UK fashion dictated a move away from casual wear to more structured garments in 1988 which inevitably hit the Hollings Mill production. In addition heavy buying by Japan, China and Eastern Europe had forced up the price of wool, making it less competitive against other fibres. Under these circumstances, it is hardly surprising that plans for a new mill were put on hold. The high value of Sterling and high interest rates further hindered a recovery in profitability and the company considered setting up an overseas shrink-treatment plant.[959]

The days when companies adorned their letterheads with pictures of their factories, the stacks pumping out plumes of smoke to emphasise productivity, were long gone and Britain in the 1980s and 1990s was far more environmentally aware. The chlorine-based shrink-proofing process produced toxic by-products and the company found itself faced with increasingly expensive anti-pollution measures.[960] Moreover, chlorine had a bad public image traceable to the German gas attacks of the First World War. In 1989 it started investigating alternative ozone shrink-proofing treatments and this became more urgent in 1991 when the Directors were informed that stricter limits were likely to be placed on adsorbable organic halides (AOX) in the effluent.[961] Thus, the company was very receptive when in 1992 Kroy suggested a three-way development of ozone treatment by Jowitts, Precision Process (Textiles) Ltd (PPT) and Kroy.[962] However, PPT made it clear that they were already working with various other concerns and that their chemical package was considerably more expensive and less foolproof than the existing process. On the other hand, the Chairman had been in discussions with Huddersfield University's Textile Department which was interested in investigating the matter.[963]

By April 1995, the company was pinning its hopes for the future on PPT's new process called EXO-S which had two major advantages: it was more environmentally friendly and it could be used prior to dyeing so that spinners could hold stocks of treated white yarn, thus eliminating unwanted stocks of dyed yarn relating to lines which were not selling well. Marks & Spencer and Coates Viyella had shown interest.[964] However, in October 1995 the new process was at least a year from being ready to run and a further year would be needed to perfect it. Even then, there were no guarantees of success. With the company facing what the Chairman called 'frightening' prospects in the coming years, W. J. B. Jowitt questioned whether the company could struggle on until the project came to fruition.[965] D. H. Foster reported in January 1996 that the company had offered to take a working machine on a trial basis with no obligation to buy. An answer was awaited from PPT and Andar, the machine's manufacturers in New Zealand. With the company in what appeared to be terminal decline, the Board planned to hive off the shrink-proofing operation into a separate company so that it could be sold more easily,[966] quite possibly to the company's dominant customer, G. Modiano Ltd. Hollings Mill Processors Ltd was duly formed and the 'hivedown' completed by the April Board meeting. Even if the company were to sell off the shrink-treatment operation, it had to continue investing in new technology such as chlorine-free processing in order to keep the plant and machinery competitive. In particular, Marks & Spencer, a very large potential customer for shrink-treated wool products, had made it clear that it was intending to move away from chlorine-treated wools within twelve to eighteen months.[967] Consequently, the company continued with the installation, in January 1997, of the new Andar machine and testing of the EXO-S process while simultaneously attempting to conclude a sale of Hollings Mill Processors Ltd to G. Modiano Ltd.[968]

Unfortunately, the possible sale of Hollings Mill Processors and the development of the new process were fraught with difficulties. In September 1997, serious problems were still being experienced with the EXO-S process. The treated tops were leaving deposits on the spinning rings, leading to yarn breakages. It was complained that PPT had come up with very few answers, and the Board agreed to suspend the hire-purchase agreement and exclusivity period until 500 kilos had been successfully treated without the unwanted deposits.[969] The machinery as well as the process was causing difficulty with consequent loss of production and a high demand on employee time. Worse still, Marks & Spencer was said to have 'hijacked' the orderly development and marketing of the process by its 'excessively enthusiastic approaches,' while at the same time failing to come through with the promised orders. At the same time, the company's hopes of a sale to G. Modiano Ltd[970] were put in doubt by the announcement in December 1998 the closure of its combing plant at Gomersal, with the expectation that it might place its business with a continental commission top shrink treater. The company had also received expressions of interest from Kroy and Bulmer & Lumb, but these, too, were thought likely to fall through as a consequence of Modiano's removal of its combing plant from England to the Czech Republic.[971]

Marks & Spencer continued to be a source of great frustration, particularly when it preferred to continue with its Italian Connection rather than what the Jowitt Board considered to be the greatly superior EXO-S garment, simply because M&S regarded the Italian label to be a selling-point.[972]

David Foster died from liver cancer in October 1999 leaving the company saddened and demoralised. Peter Bell then returned to the company as a consultant and quickly realised that the EXO-S machine made by Andar in New Zealand would never produce a viable product and that the process suffered from the fatal flaw that treated tops could not be re-treated. W. R. B. Jowitt also expressed concern about the viability of the EXO-S treatment.[973]

Woolcombers (Holdings) Ltd was the parent company of Woolcombers (Scourers) Ltd. One of its other subsidiaries, Woolcombers (Topmakers) Ltd had merged with W. & J. Whitehead (Laisterdyke) Ltd's topmaking division in November 1999. Woolcombers (Processors) Ltd closed its remaining combing mill in the same month. Unfortunately, in 2001 a receiver was appointed to Whiteheads and that mill closed.

In July 2001, Jowitts was approached by Alan J. Lewis, the Chairman of Hartley Investment Trust (Woolcombers' ultimate holding company) which led to him offering the whole of Woolcombers' shrink treatment plant to Jowitts. The plant manager, John Waddington, had already joined Jowitts in the same capacity and he now assisted in the transfer of the top treatment equipment to Hollings Mill. Jowitts purchased the top treatment line for £200,000 in October 2000. Although a loose stock line designed to treat scoured wool primarily for the Japanese futon trade was included in the price, the company declined to accept it, as Hollings Mill had insufficient space. However, given this line's origins as a top treatment line, Jowitts feared that it might be converted back and used in competition with their own top treatment plant. Rather than see the valuable machinery destroyed and therefore lost to the British wool processing industry, the company stipulated in its purchase agreement with Hartley Investment Trust that any contract for the sale of the loose stock line should include a covenant preventing its future use as a top shrink treatment line. Woolcombers (Scourers) Ltd sold the loose stock line in November 2001 and duly included the following restrictive clause (5.4) in the sales agreement:

> The purchaser warrants that for a period of 5 years from the date of Completion it shall not use the Assets as purchased to process shrink resistant wool tops.[974]

However, the £1,000 purchaser company, Speciality Processors Ltd, promptly resold the line to another £1,000 company, Precision Processors (Bradford) Ltd, this time removing the restrictive covenant and allowing the latter company to use the line for shrink treatment of wool tops, which it promptly did. Both companies were controlled by Albert Chippendale, the former manager of the Superwash Division of Woolcombers (Processors) Ltd.

Robt. Jowitt & Sons, not being a party to the sale of the loose stock line, could not itself take legal action against the purchasers, but Woolcombers commenced proceedings to enforce the terms of the sale with financial support and an affidavit from Jowitts. The litigation was expensive (approximately £80,000) and the case was far from certain to succeed; it was unclear whether clause 5.4 was enforceable since it might be considered contrary to the provisions of the Competition Act 1998 and the Treaty of the European Community, as well as constituting a restraint of trade in Common Law.[975] The case was dealt a lethal blow when Woolcombers went into voluntary liquidation on the week that it came to court. The failure of the court case cost Robt. Jowitt & Sons dear, both in the loss of the money it had lent to Woolcombers and its failure to prevent the use of the loose stock line for shrink treatment of tops in competition with Hollings Mill.

With heavy losses from the the legal proceedings, a catastrophic decline in UK combing, no continuing interest from Marks & Spencer and with Modiano now operating in the Czech Republic, there seemed to be no future for the shrink treatment facility in England. The company suffered a further blow when Alan Amos, who had started with Jowitts as an apprentice and had worked his way up to become the highly-regarded Warehouse Manager, died of a brain tumour at the age of fifty-five in 2005.

There seemed no alternative but for the company to close the mill and sell the plant to Modiano. The timing, at least, was propitious; the company had just used up the last of its tax losses on the wool business and the Sunbridge Road site was ripe for redevelopment. With perfect timing, F. T. B. Jowitt sold the mill buildings to McGinnis Developments in 2006, just before the credit crunch and slump in commercial property prices, for £2.2 million.

As always, Robt. Jowitt & Sons handled matters honourably. All affected parties were given three months' notice that processing would cease at the end of August 2006 and no orders were left unfulfilled. Every employee was given more than the statutory minimum redundancy payment. The company went into voluntary liquidation which was completed in 2007 and the proceeds distributed to the twenty-eight shareholders, mostly members of the Jowitt family.

Retrospect

Robt. Jowitt & Sons came close to extinction in 1900, flourished with an uneasy conscience during the First World War, struggled through the Great Depression and the misapplication of funds by its company secretary, made ends meet during the Second World War and rose from the ashes in the early postwar period. By the nineteen-sixties, though, it was clear that the British wool industry was struggling for survival, although R. B. Jowitt II's 1964 clarion call seemed to suggest that the problem was not terminal (p. 196). Yet there had been a slow, inexorable decline – not just in Jowitts but in the whole UK wool industry – the result of which even the most resolute action could not arrest. In one way, at least, the company had remained true to its Quaker roots by remaining solvent to the end.[976]

As has been seen, the Board of Robt. Jowitt & Sons Ltd cannot be accused of giving in without a fight. Mr F. T. B. Jowitt had recognised in 1970 that there was no future in British woolcombing (see p. 219) and in 1980 the company moved out of wool and tops altogether (p. 239), except for its involvement in specialist processing (p. 240). What was left was a very diverse, pocket-sized conglomerate, but by 2000 the Board had reverted to something more like that of the family company it had been before the purchase of S. & S. Musgrave in 1900. The only Director not descended from Frederick McC. Jowitt or (in R. J. Heaton's case) married to a descendant was the Company Secretary, was N. Matthews.[977] There were no ruthless corporate raiders like Sir James Goldsmith or Lord Hanson on the Board to carve out a new empire from the ruins of the wool company, which from the Jowitt family's financial point of view was probably a disadvantage. Where Lord Hanson's success relied heavily on slashing jobs, the Jowitts had always regarded redundancy as a last resort and R. B. Jowitt II had memorably emphasised in 1964 that 'any sacrifices to be made must first be made by Directors' (p. 196), even if his commitment to cutting his own salary was somewhat half-hearted. Holding a diversified conglomerate together, even a small one, requires not only ruthless determination and luck; it needs great depth of management experience and, perhaps above all, it needs someone like Hanson's Lord White with a nose for hidden value. Mr F. T. B. Jowitt had done his best to broaden the Jowitt Board but, with the resignation of Peter Bell in 1974, it lacked even a Managing Director. The small team was further reduced by the unfortunate death of C. M. Leveson-Gower in a car crash in November 1983. Other deaths diminished the team further.

In truth, while the diversified investments made by Robt. Jowitt & Sons were a mixed bag, they cannot in any way be regarded as being a waste of effort. Solar Water Heaters Ltd suffered from being a pioneer in its field, Bottle Recovery Ltd seems to have suffered from a failure to grasp the complexity and expense of such an operation and Access Publications was a complete disaster, although not one of Jowitts' making. Scout Alarms, with its ecclesiastical credentials, looked sure to succeed but did not. However, against these there were many successes. Jowitts made a considerable profit on the 20,000 Sirdar shares it had bought in October 1974. Peter Adams Agencies, while never quite living up to its initial promise, generated profits and handsomely repaid the initial investment when it was sold in 1978. Similarly, Everfair created substantial profits for Jowitts over a number of years and the profits from Sunbridge Electricals practically kept Jowitts afloat between 1977 and 1991. If Rutland Estates itself was not a great success, it led to the company taking a fifty percent interest in Cringles Caravan Park Ltd, yielding a large profit when it was liquidated in 1984. While it is true that Preston Gardner was not a successful investment, it is interesting to note that it was profitable when managed by Mr F. T. B. Jowitt alone in the early 1980s and returned to profit after he again took charge of it in 1984. If there was one recurring failure over the years, it was the company's reluctance to sack even the most unreliable managers. In its dealings with Victor Tate, F. M. Edwards and George Blackwell, the company could be said to have suffered from a surfeit of kindness.

Even after most of its diversified businesses had been closed or sold off, the company made every effort to renew and rejuvenate itself, bringing in a new generation – F. T. B. Jowitt's son William Robert Benson Jowitt in 1992, followed some years later by Christopher Legard, a chartered accountant who had shown great entrepreneurial flair when he founded the shirt company, Joseph Turner Ltd, in 1997, and Anthony Frieze, F. T. B. Jowitt's son-in-law and a successful banker. All these talented young people gave their time generously to the company. Sadly, however, none of them was able to devise a plan which could secure the future of Robt. Jowitt & Sons. Indeed, it was probably an impossible task.

Fig. 46 The last Chairman of Robt. Jowitt & Sons, Mr F. T. B. Jowitt, painted by Juliet Jowitt, 2011. He was President of Bradford Chamber of Commerce in 1979 and High Sheriff of West Yorkshire in 1997–8.

There was no sudden catastrophe in 2006. The company simply reached the end of its natural life, and with it died a little piece of the soul of West Riding life. The Jowitts, who for centuries defined their place in society by their involvement in the wool industry, can do so no longer. They have moved, very successfully, into new businesses but there is, perhaps, no sense of being part of some great enterprise built upon their own efforts and those of their forebears.

Appendix I: Effects of Richard Jowitt I (Transcript of the Probate Records)

An Inventory of the Goods & Chattells of Richard Jowett Late (dec[eas]ed) of Milneshay [Millshaw] within the Parish of Leeds this 23rd. day of January 1696 – appraised by us whose names are hereunto subscribed: Ro[bert] Mitchel Joshua Janson Daniel Pickering James Jubb

	£ :	s :	d
purse & Apparrel	02	00	00
In the little parlour			
One Chest of Drawers	00	15	00
One round Table	00	05	00
3 Chaires	00	03	00
2 stooles	00	01	06
5 Cushions	00	02	06
One seeing glass	00	01	00
One Drinke [ditto]	00	01	00
One Ringe	00	01	00
One Bible	00	02	06
In the great parlour			
Two bedsteads w[i]th one p[ai]r of hangings } & bedding to one bed	00	16	00
2 old cupboards	00	04	06
2 Chests	00	09	00
One Forme	00	01	00
One stoole & one Chair	00	01	06
In the Kitching			
Two Ranges & one Irone	00	08	00
One Dresser of pewter dishes & 6 plaits	01	00	00
One brass pott 2 Iron potts	02	11	00
2 pewter Tankards 2 pewter Candlesticks	00	02	06
One warming pan & 2 brass pans	00	02	06
One Chest 2 Tables	00	04	00
Four Chaires	00	02	00
Brewing pan brewing vessle & other Instr[u]m[en]ts	01	01	00
T[h]re[e] horses & furniture	09	00	00
Hay in the Barn & straw	02	00	00

251

One heiffer & 2 strikes		05	10	00
Two Items horks [?]		00	05	00

In the Milne

halfe a load of Shilling and halfe a load of blond Corn	}	00	10	00
One Bedstead & beding		00	04	06
3 Fatt piggs		04	10	00
3 holdings		01	10	00
Tot.		32	05	00
Debts Inwards		50	00	00
		82	05	00

[Also listed were his creditors:]

	£	
Isaac Blackborne	30	upon bond
Rob: Peart	30 or 40	do upon bond
Rob Hill	10	do upon bond
Jno. Atkinson	10	do upon bond
Wm Curray	8	do upon bond
his man	12	they do not know his name
and abt 200 l in Book Debts }	200	book Debts
recd one of Yorks 10£: upon bond }	10	upon bond

Appendix II: The Marriage Settlement of Rachel Crewdson

Transcribed from Cumbria Archive Centre (Kendal) WDCr/5/203/1. Such marriage settlements were a common way to protect a wife's assets in the era before the Married Woman's Property Act of 1882 and its existence does not imply any mistrust of Robert. Isaac (1780–1844) and William Dillworth Crewdson (1774–1844) were two of Rachel's brothers. Isaac was Robert Jowitt's brother-in-law, having married his sister Elizabeth in 1803. The third trustee, William Wilson (1786–1840), married Robert's sister Hannah in 1815.

𝕿𝖍𝖎𝖘 𝕴𝖓𝖉𝖊𝖓𝖙𝖚𝖗𝖊 of three Parts made the seventh day of February in the year of our Lord one thousand eight hundred and ten. 𝕭𝖊𝖙𝖜𝖊𝖊𝖓 Robert Jowitt in the County of York, Woolstapler of the first Part Rachel Crewdson of Kirkby in Kendal in the County of Westmorland Spinster of the second Part, and William Dillworth Crewdson of Kirkby in Kendal aforesaid Banker, Isaac Crewdson on Manchester in the County of Lancaster Manufacturer and William Wilson of Kirkby in Kendal aforesaid Shearman dyer and Manufacturer of the third Part, 𝖂𝖍𝖊𝖗𝖊𝖆𝖘 a Marriage hath been agreed upon and is intended, by God's Permission, to be shortly had and solemnized between the said Robert Jowitt and Rachel Crewdson, And it hath been agreed that Part of the Fortune of the said Rachel Crewdson, consisting of Nine original Shares of and in the Stocks or Funds of the United States of America, commonly called American Bank Stock, now standing in the Name of the said William Dillworth Crewdson among other Shares of his now in the said Stocks or Funds, and of the Sum of three thousand five hundred pounds three per Centum consolidated Annuities transferred into the Names of the said William Dillworth Crewdson, Isaac Crewdson and William Wilson by the said Rachel Crewdson, by and with the Consent and Approbation of the said Robert Jowitt, shall be settled transferred and assigned as hereafter mentioned. 𝕹𝖔𝖜 𝖙𝖍𝖎𝖘 𝕴𝖓𝖉𝖊𝖓𝖙𝖚𝖗𝖊 𝖜𝖎𝖙𝖓𝖊𝖘𝖘𝖊𝖙𝖍 that in Pursuance of the said Agreement, in Consideration of the said intended Marriage and of the Sum of five Shillings apiece of lawful Money of Great Britain by the said William Dillworth Crewdson, Isaac Crewdson and William Wilson in hand paid upon or before the Sealing and delivery of these Presents, the Receipt whereof is hereby acknowledged, She the said Rachel Crewdson (by and with the Privity Consent and Approbation of the said Robert Jowitt her said intended Husband testified by his being made a Party to and Sealing and delivering of these Presents) Hath Granted, Bargained, sold, assigned, transferred and set over, and by these presents Doth grant, Bargain, sell, assign, transfer and set over unto the said William Dillworth Crewdson, Isaac Crewdson and William Wilson their Executors Administators and Assigns All the said nine original Shares in the Stocks or Funds of the United States of America commonly called American Bank Stock, now standing in the Name of the said William Dillworth Crewdson and all her Right Title and Interest therein and also all her Right Title and Interest of in and the said Sum of three thousand five hundred pounds Bank three per Centum Annuities transferred to, and now standing in the Names of William Dillworth Crewdson Isaac Crewdson and William Wilson in the Books of the Governor and Company of the Bank of England and of in and to the Interest and Dividends of the same respectively. To have and to hold receive take and stand and be possessd of the same to them the said William Dillworth Crewdson Isaac Crewdson and William Wilson their Executors Administrators and Assigns Upon the several Trusts and for the several Ends Intents and Purposes hereinafter mentioned, expressed and declared of and concerning the same. And for the Considerations aforesaid, she the said Rachel Crewdson, with the like Privity Consent and Approbation of the said Robert Jowitt her said intended Husband, Hath made ordained authorized constituted and appointed and by these presents Doth make ordain authorize constitute and appoint the said William Dillworth Crewdson, Isaac Crewdson and William Wilson and each and every of them their and each and every of their Executors Administrators and Assigns her true and lawful Attornies and Attorney irrevocable for them either in her Name, or in the Names of themselves, or otherwise, but upon the Trusts and for the Intents and Purposes hereinafter mention to ask demand and receive of and from all whom it may concern the principal Sums of Money Stocks and Funds hereby assigned and the Interest and Dividends thereof respectively and to do transact and execute all such Acts Matters and Things concerning the things as fully and effectually as she the said Rachel Crewdson herself could have done in case these presents had not been made. And the said Robert Jowitt for himself his Heirs Executors Administrators and Assigns doth hereby covenant promise and agree to and with the said William Dill-worth Crewdson Isaac Crewdson and William Wilson their Executors Administrators and Assigns that the said Robert Jowitt and the said Rachel Crewdson his intended Wife shall and will at all time hereafter ratify allow and confirm all and whatsoever the said William Dillworth Crewdson, Isaac Crewdson and William Wilson shall lawfully do or cause to be done relating to the Trusts herein contained and shall not nor will revoke or make void any of the Powers or Authorities hereby in them reposed nor do any Act Matter or Thing whatsoever whereby the Receipt of the Principal Monies Stocks and Funds hereby assigned or the Interests or Dividends thereof may be impeded or delayed. And also that the said Robert Jowitt his Executors and Administrators and all and every Person and Persons whomsoever claiming or to claim by for or under the said Robert Jowitt and Rachel Crewdson or either of them shall and will at all Times hereafter do all such further Acts and Deeds if any shall be deemed necessary which the said Trustees may require for the better assigning the said Stocks and Funds and Interest Right and Title of the said Rachel Crewdson therein. And this Indenture further witnesseth that for the consideration aforesaid, it is hereby declared and agreed by and between the said Parties to these presents, that the said William Dillworth Crewdson, Isaac Crewdson and William Wilson their Executors Administrators and Assigns, shall stand and be possessed of, interested in and entitled to as well the said nine Shares of American Bank Stock as of the said three thousand five hundred consolidated three per Centum Bank Annuities and the Interest Dividends and Proceeds thereof In Trust for the said Rachel Crewdson and her Assigns until the said intended Marriage shall be had and solemnized. And from and after the Solemnization thereof. Upon Trust that they the said William Dillworth Crewdson, Isaac Crewdson and William Wilson and the Survivors and Survivor of them, and the Executors or Administrators of such survivor do and shall at the Request, and with the Consent in writing of the said Rachel Crewdson, notwithstanding her Coverture, sell, dispose of and transfer the said American Bank Stock and three thousand five hundred pounds three per Centum consolidated Bank Annuities and lay out and invest the Money thence arising in the Names or Name of the said Trustees or the Trustees or Trustee for the Time being, either in the public Funds or Parliamentary Stocks of Great Britain or of the United States of America at Interest upon government or apparent good, real or personal Securities in Great

Britain, and by such Consent as aforesaid, alter, vary, and transpose the said Stocks, Funds and Securities for others of the like Nature, as often as they shall be requested to do so by the said Rachel Crewdson. And the said William Dillworth Crewdson, Isaac Crewdson and William Wilson and the survivor of them his Executors Administrators and Assigns shall stand and be possessed of and interested in such Stocks Funds and Securities as aforesaid. Upon Trust to pay the Interest Dividends and annual Proceeds of such Stocks Funds and Securities unto the said Rachel Crewdson during her natural Life for her own Use and Benefit free from the Debts Engagement or Control of her said intended Husband, and for which her Receipt alone, notwithstanding her Coverture, shall be to them a good and sufficient discharge. And from and after the Decease of the said Rachel Crewdson, Upon Trust to pay and transfer such Stocks Funds and Securities and the Interest Dividends and annual Proceeds that shall be then due on the same, unto such Person or Persons, at such Time and Times and in such Proportions Manner and Form, and subject to such Conditions and Limitations over, as the said Rachel Crewdson shall by any Deed or Deeds, Writing or Writings two or more credible Witnesses to be by her sealed and delivered in the presence of, and attested to by two or more credible Witnesses, or by her last Will and Testament in Writing to be by her signed sealed published and declared in the presence of and attested by two or more credible Witnesses direct or appoint. And in default of such Direction or Appointment, and as to such Part or Parts of the said Stocks Funds and Securities of which no such Direction or Appointment shall be made Upon Trust to pay and transfer the same unto such Person and Persons as shall be the next of Kin of the said Rachel Crewdson according to the Statute for the Distribution of Intestate Estates, and to for and upon no other Trust Intent or Purpose whatsoever. Provided always and it is hereby declared and agreed that the said William Dillworth Crewdson, Isaac Crewdson and William Wilson and the Survivors and Survivor of them and his Executors and Administrators shall not be answerable or accountable for any Loss that may happen by lending the said Trust Monies upon Interest or in lodging the same in any Bank or Bankers hands for safe Custody such Loss happening without their or his wilful Neglect or default, nor shall they or any of them be answerable or accountable for the Acts, Receipts, Neglects or Defaults of each other but only each of them for his own Acts, Receipts, Neglects or Defaults. And that they and each of them shall be allowed to deduct and retain to themselves all their and each of their Costs and Charges for Loss of Time and Expences, as they shall respectively suffer, pay, expend or be put unto, in the Execution of the Trusts hereby in them reposed, from and out of the said Trust Monies. In witness whereof the said Parties to these presents have hereunto set their Hands and Seals the day and Year first within written.

Signed sealed and delivered being   Robert Jowitt
first duly stamped in the presence of us }   Rachel Crewdson
     John Jowitt   Wm Dillworth Crewdson
     Isaac Wilson   Isaac Crewdson
     Cicely Crewdson   Wiliam Wilson

Appendix III: Branch Profit/Loss Accounts 1899–1908 (Incomplete)

William Gibson & Co. profit/loss to nearest £							
Year	1899	1900	1901	1902	1903	1904	1905
Profit (Loss)	£14,647	(£10,801)	£5, 410	£9,062	£5,688	£4,765	£44,52

East london Branch profit/loss to nearest £							
Year	1901	1902	1903	1904	1905	1906	1907
Profit (Loss)	£115	£8,492	£4,101	£228	(£603)	(£2186)	(£1620)

Port Elizabeth Branch profit/loss to nearest £							
Year	1901	1902	1903	1904	1905	1906	1907
Profit (Loss)	£326	£2874	£529	£10	£318	£419	N/A

West of England Branch profit/loss to nearest £							
Year	1901	1902	1903	1904	1905	1906	1907
Profit (Loss)	£588	£1553	£761	£335	£70	£244	N/A

Melbourne Branch profit/loss to nearest £								
Year	1901	1902	1903	1904	1905	1906	1907	1908
Profit (Loss)	(£23)	£2704	(£96)	(£434)	N/A	N/A	N/A	(£637)

Bradford Branch profit/loss to nearest £				
Year	1904	1905	1906	1907
Profit (Loss)	£5940	£6813	£8045	£9075

Appendix IV: RJ&S and Predecessor Partnerships Profit/Loss 1807–1902

(1889, 1890, 1899 missing)

Year	Profit £	Loss £	Year	Profit £	Loss £	Year	Profit £	Loss £	Year	Profit £	Loss £
1807/8	1211		1834	1164		1857	3384		1880	127	
1809/10	4374		1835	761		1858		1723	1881	1442	
1811/12		507	1836	2619		1859	3462		1882	617	
1813	699		1837		6242	1860	3700		1883	2203	
1814	3671		1838	1760		1861		765	1884	3644	
1815	1495		1839	1758		1862		909	1885	254	
1816		1139	1840		904	1863	4386		1886	12223	
1817	2184		1841	1497		1864	3178		1887	4892	
1818	3618		1842	1866		1865	4993		1888	10340	
1819		2844	1843	1328		1866	5288		1889		
1820		2518	1844	5528		1867		5467	1890		
1821		215	1845	7559		1868		989	1891		1549
1822	820		1846	1478		1869	2670		1892		8128
1823		12	1847	3564		1870	3287		1893	3442	
1824	840		1848		997	1871	20007		1894		6221
1825	3447		1849	6270		1872	4137		1895	6718	
1826		8436	1850	6238		1873	5681		1896	4895	
1827		200	1851	3630		1874	11308		1897	2630	
1828	748		1852	4543		1875	7125		1898	328	
1829	301		1853	4909		1876	5500		1899		
1830	2283		1854		950	1877	5284		1900		74102
1831	3297		1855	2538		1878	1708		1901	39	
1832		105	1856	3043		1879	1905		1902	15336	
1833	5086										

Appendix V: Initial Share Allocations

Taken from Peter Bell, *Robert Jowitt & Sons, the Limited Company.*

80,000	Frederick McCulloch Jowitt, Permanent Director, Chairman 1919–21 (d. 1921)
36,800	Edward Maurice Jowitt, Permanent Director, Chairman 1921–38 (d. 1954)
24,900	Robert Jowitt, Permanent Director (d. 1945)
15,000	Mrs Helen Dorothea Jowitt (d. 1952)
11,800	to F. McC., E. M. and R. Jowitt as executors of Robert Benson Jowitt (d. 1914)
168,500	*Shares allocated as fully paid*

2	F. McC. and E. M. Jowitt as subscribers to the Memorandum of Association

22,001	Sam Harland, Permanent Director 1919–54, Managing Director 1919–39 (d. 1954)
5,000	Herbert Lee, Ordinary Director (d. 1925, shares to his daughter, Mrs E. M. Tyne)
2,497	Captain F. R. B. Jowitt, Ordinary Director 19941, Chairman 1948–58 (d. 1965)
2,000	George Blackwell, Ordinary Director and Secretary (dismissed 1931)
31,498	*Allotted for cash*

200,000	*Total number of shares issued in 1919*

In 1923 the Board capitalised £50,000 from the General Reserve Fund, allotting the shares *pro rata* to the existing shareholders:

19,733	A. T. Keeling, G. C. Veale & G. Blackwell as trustees of estate of F. McC. Jowitt
10,183	E. M. Jowitt
7,208	Robert Jowitt
3,750	Helen Dorothea Jowitt
6,750	Sam Harland
1,250	Herbert Lee (d. 1925, shares transferred to Mrs E. M. Tyne)
624	Captain F. R. B. Jowitt
500	George Blackwell
2	Sam Harland (initial subscription shares)
50,000	

Appendix VI: Salaries Agreed on 27 October 1920

Source: Peter Bell, *Rbt. Jowitt & Sons, the limited company,* and minute books.

	per annum	
Frederick McCulloch Jowitt	£1,000	
Sam Harland	£1,500	+ 2½% commission on trading profit
Herbert Lee	£1,400	+ 2½% commission on trading profit
George Blackwell	£900	+ 1¼% commission on trading profit
Donald B Sykes	£750	+ commission on Port Elizabeth profit
G. A. Reid	£750	+ commission on East London profit
Arthur Gibson	£750	+ commission on Australian profit
Joe Sheard	£400	
H. C. Edwards	£400	
Capt. F. R. B. Jowitt	£300	

Appendix VII: Top Department Wages Bill 1915–20

The following figures are taken from Ledger K (Brotherton 73, Hudson 72) which contains records for the Top, Waste, London Office, Central Haulage and De-wooling departments. These figures are for the year to 20 November.

	Total	Warehousemen	Sorters	Foremen
1915	£6,116 10s 3d	£732 3s 7d	£4,037 1s 4d	£1,347 5s 4d
1916	£4,891 18s 11d	£745 15s 11d	£2,963 12s 1d	£1,182 10s 11d
1917	£6,985 11s 11d	£1,001 10s 8d	£4,597 0s 2d	£1,359 5s 0d
1918	£9,262 2s 2d	£1,049 11s 11d	£7,844 2s 6d	£91 3s 5d
1919	£2,011 10s 9d	£836 3s 1d	£1,122 11s 11d	£0
1920	£3,483 14s 11d	£2,578 19s 6d	£850 8s 1d	£0

Appendix VIII: Top Department Insurance Costs 1916–20

The following figures are taken from Ledger K and are for the year to mid-November.

1916	£21,041 5s 1d*
1917	£35,560 9s 0d
1918	£7,710
1919	£797 19s 3d
1920	£2,108 19s 7d

* Consisting of £1,798 19s 8d fire; £17,600 18s 3d marine; £42 11s 3d compensation; and a surprising £1,618 15s 11d aircraft.

Appendix IX: Keeling's Memorandum on Turning Robt. Jowitt & Sons into a Limited Company, 7 December 1910

The method of changing your business into a private limited company would shortly be as follows:

There would be a short written Agreement under which the partners in the present firm would sell the business to a limited company for an agreed price which would be satisfied by the allotment and issue to the partners of fully paid up Ordinary shares of an amount to be agreed.

The fully-paid shares will be issued to the existing Partners in the ratio in which they now take the profits of the business.

Each of the partners would be required to enter into a covenant that he would not carry on or be in any way involved in the business of a wool broker or use or allow his name to be used in connection with any such business within a limited number of miles from the various places at which the Company will carry on its business.

The private Company may be formed by two or more persons and you must by the articles restrict the right of transfer of the shares and you can never have more than fifty shareholders and you cannot invite the public to subscribe for shares in the Company.

A private Company has not to include in its annual summary and return any balance sheet. A form of return is enclosed from which it does not seem, as page 2 is excluded, there is anything to which you could object.

The working capital of the Company would be provided by the issue of say £80,000 5% Cumulative Preference shares, which should be preferred both as regards to capital and income, and these shares would be taken up by the partners out of the capital in the existing business.

The chief difficulty which arises in connection with a private limited company of this kind is when a death occurs.

It is obvious that the power to transfer the shares must be restricted to shareholders in the Company, as otherwise your competitors in business might acquire a holding and interfere in the management of the Company, but, on the other hand, provisions must be inserted which would prevent a partner retiring and still continuing to draw the dividends of his shares which would be carried for him entirely by the active members in the business.

In the case of death, too, it is usual to provide that a maximum dividend should only be paid on the holding of a deceased shareholder until his sons or other nominees are in a position to come into the business, and in the meantime any dividend over and above the amount fixed goes to the other shareholders in the Company and increases their dividend.

The directors are by the memorandum and articles of association bound to admit the partners' nominees, subject to this reservation, that any nominee is in the opinion of the directors undesirable, the question of desirability of admission is to be submitted to the decision of some independent person to be nominated jointly by the directors of the Company and the Executors of the deceased partner.

The remuneration of the directors must be fixed at a maximum amount by the memorandum and articles of association [and] should not be altered. The voting power must be adjusted so that each share will carry one vote only for the purpose of voting in favour of any proposal, but for the purpose of voting against any proposition to alter the Memorandum and Articles of Association of the Company, each share will carry 200 votes.

The object of this is to prevent the Directors being in a position to vote themselves practically unlimited remuneration and thus destroy the value of the ordinary shares.

You need not consider the question of your various employees to whom you now pay commission on the profits. They can if they wish it all invest their capital in preference shares of the new Company which will pay 5% and the Directors will have power to grant them salaries with a commission ranging with the profits.

Practically, the directors will have an absolutely free hand for all salaries and remunerations except their own.

The first question, of course, is the bank and the bank may object to your turning yourselves into a limited Company but this does not seem an insuperable objection as it is always open to the Directors to give the Bank a joint and several guarantee for the amount of the Company's overdraft, but care must be taken in giving this guarantee to limit any partner's liability to 6 months after the date of his death.

This course is preferable to leaving shares partially, as the executors of a deceased partner would not like to hold partially paid shares, when practically the whole of the partners' fortune is in the business.

The number of ordinary shares ought to be based on a return of 10%.

The real question that you have to settle is suppose you or one of the partners die what would be a fair sum for the widow and children to draw from the business pending the deceased partner's son coming into the business and is a deceased partner's son eventually to succeed on his coming into the business to the whole of the share of profits of his father.

It is possible to provide by the Memorandum and Articles of Association for practically any arrangement that is agreed upon and such an arrangement can in effect be made irrevocable by adjusting the voting power as above mentioned. As however the arrangement is in effect irrevocable it is most important that the Memorandum and Articles of Association should be very fully considered from every point of view as the arrangements provided for by the Memorandum and Articles of Association are to all intents and purposes final.

If there are any suggestions which may occur to you on reading this perhaps you will note them in the margin as it would be well if the matter is going on that some intelligible scheme should be put before the Partners at the New Year for consideration.

The matter can best be considered by preparing a form of Memorandum and Articles of Association and the partners making their criticisms on the draft.

Appendix X: Robt. Jowitt & Sons
AGM Financial Statements 1919–67

Trading profit shown after payment of Directors' Fees, Management Salaries, Commissions, Bonuses and Tax. Charges and Tax taken directly from Reserves are not shown. The trading profit as defined in the AGM statements sometimes differs considerably from the trading figures given in Board meetings.

Source: Peter Bell, *Robert Jowitt & Sons, the Limited Company*

Year ending 20 Nov.	Trading profit plus Brought forward £k	Life policy premiums (increase in values) £k	Transferred to reserves (from reserves) £k	Dividends paid Ordinary *Preference £k	Balance carried forward £k	Share valued declared at AGM £k
(6 months) 1919	85.4	1.8	74.0	8.0	1.6	
1920	157.8	1.9	20.0 Tax 95.0	20.0	0.9	22/6
1921	106.0		45.0 Tax 7.0	20.0	0.6	25s
1922	78.4		50.0	20.0	8.4	25s
1923	56.7		27.0	25.0	4.7	25s
1924	56.7		25.0	25.0	(6.7)	25s
1925	(67.4)		(68.4)		1.0	24s
1926	31.7		10.0	18.8	2.9	21s
1927	40.2		15.0	25.0	0.2	20s
1928	4.8	(7.7)		12.5	0.2	22s
1929	(31.2)		(44.0)	12.5	0.3	21s
1930	(29.3)		(21.6)		(7.7)	17s 6d
1931	(34.3)				(34.3)	15s
1932	4.1			2.5	1.6	12s
1933	60.2		40.0	20.0	0.2	15s
1934	(49.9)		(40.0)		(9.7)	17/6
1935	49.6	(2.9)		20.0	29.6	13/6
1936	63.6	(0.3)	15.0	12.5	36.4	17/6

Year ending 20 Nov.	Trading profit plus Brought forward £k	Life policy premiums (increase in values) £k	Transferred to reserves (from reserves) £k	Dividends paid Ordinary *Preference £k	Balance carried forward £k	Share valued declared at AGM £k
1937	(64.3)		(25.0)	6.3	(9.2)	17/6
1938	20.9	(0.6)		6.2	6.1	14s
1939	34.3	(0.2)		18.8	15.5	15s
1940	42.3	(1.1)		18.7	14.3	16s 6d
1941	16.6	(0.2)		15.0	42.0	20s
1942	19.4	(0.2)		18.8	42.8	20s
1943	13.4	(0.2)		12.5	43.9	20s
1944	15.6	(0.2)	0.4	15.0	44.3	20s
1945	19.3		46.8	15.0	1.8	22s 6d
1946	100.4		71.0	25.0	4.4	24s
1947	160.8		127.5	25.0	8.3	40s
1948	460.2		176.2	25.0	12.2	42s 6d
1949	160.6		112.5	25.0	23.1	45s
1950	278.5		30.0	30.0	25.3	No value
		Stock	150.0	*3.9		declared
		Pension	39.3			after 1949
		scheme				
1951	89.9			22.5	61.1	
				*6.3		
1952	160.9		50.0	45.0	59.6	
				*6.6		
1953	285.1		180.0	37.1	61.4	
				*6.6		
1954	97.1	(5.5)		37.1	58.9	
				*6.6		
1955	28.0		(30.0)	30.2		
			(7.5)	*6.6		
				7.5		
1956	119.0		30.0	38.8	43.3	
				*6.9		
1957	(104.1)		(170.0)	30.2	28.8	
				*6.9		
1958	(55.2)		(130.0)	*6.9	67.9	

Year ending 20 Nov.	Trading profit plus Brought forward £k	Life policy premiums (increase in values) £k	Trans-ferred to reserves (from reserves) £k	Dividends paid Ordinary *Prefer-ence £k	Balance carried forward £k	Share valued declared at AGM £k
1959	185.4		75.0	31.0 *7.4	72.0	
1960	(28.5)	Adelaide sale surplus	(75.0) (28.6)	*7.3	67.8	
1961	108.6			14.7 *7.4	86.5	
1962	74.8			9.2 *7.3	58.3	
1963	126.8			18.3 *7.4	91.1	
1964	(13.00)	Subsidiary loss	(30.0) 6.9	*7.3	16.6	
1965	16.6		(252.4)	*7.2	4.2	
1966	25.0	Subsidiary profit	(3.2)	gross *12.0	9.8	
1967	(44.6)		(60.0)	gross *12.0	3.4	

Appendix XI: Memorandum of Proposed Agreement as to the Business of Walter Scott & Sons

1.   There shall become temporarily dormant the indebtedness (a) to Robt. Jowitt & Sons Ltd. of approximately £80,000, and (b) to the Bank of approximately £60,000, subject to the proviso that Messrs Scott shall be entitled to increase said indebtedness to the Bank from time to time up to £65,000 by making payments-out of wages, costs of dye wares and other outgoings connected with the Joint Adventure between Messrs Scott and Messrs Jowitt as constituted by letters interchanged between them, this Bank Account shall be denominated No. 1 Account for the purposes of this memorandum.

2.   Messrs Scott shall open with the Bank a No. 2 Account which shall be used for general purposes (but excluding said Joint Adventure), and this account may be over-drawn up to the maximum limit of £10,000, and Messrs Jowitt shall be liable to the Bank for one half of the debit balance on this Account.

3.   There shall also be opened a Joint Adventure Bank Account in the name of Messrs Scott and Messrs Jowitt, and there shall be placed to the credit of this Account the proceeds of Sales of Joint Adventure materials, and such proceeds shall from time to time be divided in proportion to the contributions of the respective parties to the Joint Adventure as far as is required to reimburse the same, but any balance representing profit shall not be divisible except by consent of the two parties to the Joint Adventure.

Appendix XII: Combing Charges (Tariff) 1933, 1937 and 1990

Source: Peter Bell, *History of the Woolcombers Mutual Association Limited* (after *Weekly Wool Chart* vol. 31 no. 1580)

Noble Combing per lb Top & Noil			French Combing per lb (1990 prices per kg) Top & Noil				
	1933	1937		1933	1937		1990
Merino qualities			Merino qualities			*70's*	86 p
	3.70 d	4.10 d	*Tear 4/1 & over*	3.95 d	4.70 d	*64's*	84 p
			Tear under 4/1			*60's*	82 p
Crossbred carding			Crossbred carding				
58's–56's	3.15 d	3.50 d	*58's–56's*	3.60 d	4.70 d		73 p
50's & coarser	2.75 d	3.10 d	*50's & coarser*	3.60 d	4.70 d		71 p
Prepared							
48's & finer	2.30 d	2.75 d	*Note: the 1937 French combing prices*				
46's & coarser	2.30 d	2.60 d	*are subject to a 10% conditional rebate*				

Appendix XIII: Some Descendants of Joseph Jowitt Senior (1757–1803)

As noted in the main text, Joseph Jowitt and Samuel Birchall split their business from that of Joseph's brother, John Jowitt II in 1802. Joseph died in the following year. Joseph's sons, Thomas, John, usually referred to as John Jowitt Sr, and Joseph Firth Jowitt (Joseph jr), continued in business as wool staplers in Leeds until 30 December 1837 when Thomas and Joseph Jowitt Jr retired from the business.

EDWARD JOWITT OF ELTOFTS (1806–49)

Thomas Jowitt was clearly a wealthy man because his son Edward Jowitt of Eltofts, lived a very comfortable life as a country gentleman and justice of the peace. A description of his estate in 1849 gives a clear picture of his lifestyle:

THE MANSION & ESTATE called ELTOFTS, the residence of the late Edward Jowitt, Esquire, in the parish of Thorner, in the West-Riding of the county of York, and comprising 151A, 2R. 3P. [i.e. over 151 acres] of superior LAND, in the highest state of cultivation. The Mansion is suitable for the residence of a large genteel family, containing dining-room, 27 feet by 18; drawing-room and breakfast-rooms, each 24 feet by 10; and a smaller room, with servants' hall, butler's pantry, store-rooms, &c., all replete with fixtures on the ground floor; six good bed-rooms, two dressing-rooms, and ample servants' rooms, two water closets, and very convenient day and night nurseries; large commodious kitchens, wine and ale cellars; also near the Mansion, detached laundry, brewhouse, and other offices, billiard-room, spacious double coachhouse, saddle and harness rooms, and stabling for nine horses, exclusive of several loose boxes. Also, excellent granary, barn, strawfold with sheds, cowhouse, thrashing machine and grinding mill, attached dove cote and dog kennels.

There are also, at a convenient distance, a hind's house, and three good cottages, barn, two cart-horse stables, beast houses, sheds, strawfolds, piggeries, blacksmith's shop, and other farm buildings and conveniences of a superior description, adapted to the occupation of the land, with excellent water.

♂ **Joseph Jowitt sr**
b. 01 Oct 1757 at Pudsey
m. 30 May 1781 Grace Firth (1758 -
   1846)
d. 17 Mar 1803 at Leeds, aged 45

♀ **Mary Jowitt**
b. 08 Jun 1782 at Churwell
d. 30 Sep 1783, aged 1

♂ **Thomas Jowitt**
b. 10 Feb 1784 at Churwell
+. Mary Walker
d. 11 Mar 1851 at Chapel Allerton, aged
   67

♀ **Ann Jowitt**
b. 08 Feb 1786 at Leeds
d. 20 Apr 1865 at Toothill, aged 79

♀ **Sarah Jowitt**
b. 15 Oct 1787 at Leeds
d. 20 Nov 1824, aged 37

♂ **John Jowitt Sr**
b. 1790 at Leeds
m. 29 Jan 1829 Mary Ann Norton
   (1804 - 1883)
d. 1860 at Leeds, aged 70

♀ **Grace Jowitt**
b. 29 Nov 1792 at Leeds
d. 22 Dec 1870 at Leeds, aged 78

♀ **Maria Jowitt**
b. 27 Feb 1795 at Leeds
m. 01 Apr 1816 Robert Arthington
   (1779 - 1864)
d. 05 Oct 1863 at Leeds, aged 68

♂ **Joseph Firth Jowitt**
b. 22 May 1799 at Leeds
m(1) 23 Oct 1828 Mary Lupton (- 1852)
m(2) 06 Jul 1854 Hannah Ellis
d. 05 Apr 1863, aged 63

The whole premises are in complete repair; extensive flower and kitchen gardens, well stocked with choice flowering shrubs, evergreens and flowers, with fruit trees in full bearing, adjoin the Mansion, and tastefully arranged plantations and timber trees ornament the estate. Eltofts is situated near the turnpike road from Leeds to Wetherby, from each of which it is distant about six miles, and four miles from Harewood.

The estate abounds with game, being amongst preserves; and shooting, to almost any extent, may be obtained in the vicinity; a trout stream forms the northern boundary; the Bramham Moor Foxhounds hunt the country; the kennels are only four miles off, and the greater part of the meets within a few miles.[978]

The life of a country squire could have its dangers, though. On 21 October 1835, John jr recorded in his diary,

Poor Edward Jowitt was almost killed to-day, hunting! His horse fell, in leaping a blind ditch, & struck him on the head; & he was carried, on a gate, to Becca Hall near Bramham. 'Like a slain deer the tumbril brings him home'![979] – poor, poor fellow! Gay, rich, handsome, young & amicable – alas alas! – his poor soul – his poor soul. May God spare his life – oh may He, but if not, oh give him consciousness & grace to repent & believe before he dies! Poor Cousin Mary Ann [John Jowitt senior's wife Mary Ann, so Edward's aunt] seemed very much disturbed.[980]

John jr's concern for his soul does not necessarily indicate that he thought that Edward was particularly wicked. John was a deeply religious man and was frequently reproving himself for what he considered lapses of faith.

Edward died in 1849 at the age of forty-three and his estate was put up for sale.

ROBERT ARTHINGTON (1823–1900)

Despite coming from a family which contained many members of the most temperate habits, Joseph senior's daughter Maria married Robert Arthington, a member of the wealthy Leeds brewing family on 1 April 1816.[981] She was not the first member of the family to marry an Arthington. John Jowitt I's brother, Benjamin, married an Ann Arthington in

1815 and their son, Benjamin Jowitt of Carleton, was close to the Jowitts of Leeds.

Maria was an eminent Quaker preacher and author of many booklets, including *A Few Remarks on the Leading Principles of Christian Faith, The Little Scholar's First Grammar; or Grammar Made Easy to the Capacities of Young Children, and Rendered Pleasing by a Variety of Familiar Examples; Adapted to the Use of Private Families,* and *Queries for Women Friends.* All but one of an album of fifty silhouettes in the Brotherton Library, University of Leeds, are believed to be by her. In 1861 Maria and Robert were living at 46 Hunslet Road, while Maria's sister Grace, a spinster, lived at number 44. Robert and Maria's son, also Robert, was undoubtedly the most eccentric of the Jowitts' relations, and one of the wealthiest. On the death of his father in 1864, he inherited £200,000, investing well and spending little. His entries in the 1871 and 1881 censuses describe him as 'Consul at Leeds for the Republic of Liberia'. This might seem fanciful but the appointment is confirmed by the *London Gazette,* 2 April 1869. Arthington is known to have given money to help former American slaves settle in Liberia and there is a town there named after him.

Until moving to Teignmouth for his health in about 1896, he spent nearly the whole of his adult life in a large villa in a Leeds suburb[982] which was 'scarcely, if ever, entered by a woman, and, consequently, scarcely, if ever, swept'.[983] Not only was the house unswept, it was very full of papers:

One of the eccentricities of the Arthington family, and the late Mr. Arthington inherited it perhaps in an intensified form, was the care with which they treasured old documents, manuscripts, albums, letters, account books, writing books and other school exercise books dating back more than two centuries. Nothing, not even the meanest pamphlets, appear to have been destroyed; thus the labours of the executors in wading through the tons of printed and written matter which has accumulated may be faintly imagined. As may be expected, the books and pamphlets on missionary and evangelisation matters, and the translations of the Bible into *foreign* tongues, form a larger proportion of the later literature, but among much that will have to be consigned to the flames there are some curiosities which are well worth referring to, and some books which are really valuable. Among the former may be classed the following: – Collection of primitive spectacles, quantity of early steel pens, family account books from 1670, large collection silver buckles as worn last century on shoes and knee breeches; large number of pocket-knives belonging to members of the family for the last 150 years; family trinkets and miniatures, beautifully executed; several old watchman's rattles; some very fine engravings by Albert Durer; and a writing copybook in various hands (1573), bound in parchment. Among an immense quantity of books are a first edition of Penn's "No Cross, No Crown;" Dr. Johnson's works, nine volumes; Defoe's "Robinson Crusoe," with map, early edition; Barclay's "Apology," sumptuously bound, presented by a grandson of the author; "The Dance of Death," pictures by Hans Holbein, engraved by W. Hollar, two volumes (valuable); another edition by T. Rowlandson. Probably the oldest book in the collection is a copy of the New Testament, issued by Erasmus about 1510. There are also very fine copies of Thoresby's "Leeds," and Dugdale's "History of the Late Troubles in England."[984]

His Teignmouth retreat was said to be 'cleaner, if humbler,' for, on arriving in that town, he asked an old boatman for lodgings:

The boatman, seeing an aged man of poverty-stricken appearance, offered him quarters in his own house. And there Robert Arthington ended his days among kindly people, who had no suspicion of his fabulous wealth.[985]

One obituary described Arthington thus:

When a resident of Leeds, he used to walk about the streets, a puny, shrivelled figure, dressed in a suit of no fashionable cut, grasping a stout umbrella and his little head was surmounted by a silk hat several sizes too big. He resided in Headingley-lane,[986] the house standing in its own grounds, and the deceased gentleman was very wealthy. He lived carefully, and some people called him a miser, but he was not so. There was one luxury that he permitted himself – the luxury of contributing to missionary enterprise.[987]

Arthington had been a generous contributor to missionary and other charities during his lifetime, sometimes, as in the case of £1,000 each donated to Indian famine relief funds in Leeds and London, under the pseudonym 'Thine own I have given Thee'.[988] The main object of his charitable donations were the Baptist Missionary Society (BMS) and the London Missionary Society (LMS). The BMS Congo mission was founded by his donation of £1,000 in 1877 and the Baptist missionary George Grenfell's riverboat, *Peace,* was financed by a further donation of £4,000 in 1880, supplemented by another £1,000 in 1882. He gave the money on condition that the BMS advanced up the Congo river to meet a possible extension of the LMS Tanganyika mission (also financed by Arthington).[989] Henry Stanley asked Arthington for permission to use the boat in his expedition to rescue Emin Pasha. Arthington declined.

Shortly before his death he gave £20,000 through his friend Mr John Edmund Whiting to Robert Benson Jowitt, as Treasurer of Leeds General Infirmary, and Henry Barran, a member of the committee of the Hospital for Women and Children. They were to put £8,000 towards rebuilding the latter institution and use £12,000 to erect 'a Semi-Convalescent hospital for Women in connection with the Infirmary'.[990]

After his death in 1900, his estate was estimated at between £900,000 and £1,000,000[991] but the will was poorly drafted and by the time it had been sorted out in 1910 the value had risen to £1,273,849. Only ten percent of this was left to his relations, and that only at the insistence of his friend, John Edmund Whiting.[992] Fifty percent of the estate went to the BMS and the other forty percent to the LMS. Arthington has been called 'possibly the most important Protestant missionary strategist of the nineteenth century,'[993] but his influence went well beyond Africa. He was particularly interested in India and china and he founded the Arthington Aborigines Mission at Mizoram in north-east India in 1894.[994]

JOHN JOWITT SENIOR (1790–1860)

John Jowitt senior married Mary Ann, daughter of Thomas Norton, at the Friends' Meeting House in Peckham Rye on 29 January 1829. The wool trade seems to have made him a reasonably wealthy man, just as it did his brothers. The 1851 census shows him living in retirement, surrounded by his family, at 10 York Place, Leeds. It was a prosperous but unpretentious area populated by surveyors of taxes, dyers, stuff merchants and druggists. He had two daughters, Elizabeth and Marianna, and seven sons, John Henry, William, Joseph Firth, Alfred Norton, Frederick, Thomas Norton and Walter Edward. Alfred emigrated to New Zealand and Thomas to Australia, where a number of his descendants still live.

WILLIAM JOWITT (1834–1912)

William received his BA from University College Durham on 11 June 1858.[995] He left the Society of Friends in 1859 and received his MA in 1862. He was ordained by the Bishop of London in the following year and was immediately appointed to the curacy of St Thomas, Charter-house.[996] In August 1866, the Corporation for Middle Class Education in the Metropolis and the Suburbs appointed him the headmaster of their first school,[997] a role in which he excelled. In the same year he married Lousia Margaret Allen (1841–1920), third daughter of John Allen of Old-field Hall, Altrincham, Cheshire. In 1874 he became Rector of Stevenage. He had nine daughters, Dora, Margaret, Grace, Kathleen, Mary, Ruth, Lettice, Evelyn, Audrey and only one son, William Allen.

WILLIAM ALLEN JOWITT (1885–1957)

From the age of seven, William Allen Jowitt attended Northaw Place School in Potters Bar, where the future Labour Prime Minister, Clement Attlee, was a contemporary. In January 1899 he was sent to Marlborough College, which his cousins Frederick McC. Jowitt and Edward Maurice Jowitt, as well as A. T. Keeling, had attended some years before. He entered New College, Oxford, where he gained a first-class honours degree in Jurisprudence in 1906. He was admitted to the Middle Temple in 1906, becoming a pupil of A. J. Ashton K. C., and was called to the Bar in 1909. It is clear, however, that like so many other barristers his true vocation was not the law but politics. He started working for the Liberal Party during the 1910 election. In 1913 he married Lesley, second daughter of J. P. McIntyre.

He joined the Royal Naval Volunteer Reserve on 20 October 1914 but resigned on 18 November due to ill health.[998] His enemies said that his lack of military service gave him a chance to build up his practice.

He was elected as the Liberal MP for Hartlepool in 1922 but lost the seat in the 1924 election. On 30 May 1929 he was returned as one of the Liberal members for Preston. His friend Ramsay MacDonald then invited him to join the Labour Party and become Attorney General, which he did, receiving the customary knighthood. This switch in alliance led to bitter disputes about his integrity in legal and political circles, prompting him to resign and stand in the ensuing by-election which he won by a majority of 6,440 votes. When Ramsay MacDonald's faction of the Labour Party, the National Labour Organisation, joined the National Government of 1931, Jowitt went with it and was expelled from the official Labour Party. In the general election of October that year, he stood as National Labour candidate for the Combined Universities. As he had

♂ **John Jowitt Sr**
b. 1790 at Leeds
m. 29 Jan 1829 Mary Ann Norton
  (1804 - 1883)
d. 1860 at Leeds, aged 70

♀ **Elizabeth Jowitt**
b. 09 Dec 1829 at Leeds
d. 1885 at Manchester, aged 56

♂ **John Henry Jowitt**
b. 07 Dec 1832 at Leeds
m. 14 Aug 1866 Helen Monro

♂ **William Jowitt**
b. 02 Jul 1834 at Leeds
+. Louise Margaret Allen (1841 - 1920)
d. 09 May 1912 at Stevenage, aged 77

♂ **Joseph Firth Jowitt**
b. 08 Feb 1836 at Leeds

♀ **Marianna Jowitt**
b. 07 Jul 1837
d. 19 Jul 1865 at Manchester, aged 28

♂ **Alfred Norton Jowitt**
b. 17 Apr 1839 at Leeds

♂ **Frederick Jowitt**
b. 05 Feb 1841 at Leeds
d. 04 Mar 1846, aged 5

♂ **Thomas Norton Jowitt**
b. 11 Jul 1841 at Leeds

♂ **Walter Edward Jowitt**
b. 08 Oct 1843 at Leeds
m(1) 11 Jul 1877 Gertrude Birchall (-
  1878)
m(2) 29 Jun 1880 Mary Arthur
  Mathieson (- 1888)

♀ **Dora Jowitt**
b. 1867 at Finsbury

♀ **Margaret Jowitt**
b. 1869 at Finsbury

♀ **Grace Jowitt**
b. 1873 at Islington

♀ **Kathleen Jowitt**
b. 1874 at Navestock, Essex

♀ **Mary Jowitt**
b. 1875

♀ **Ruth Jowitt**
b. 1877 at Stevenage

♀ **Lettice Jowitt**
b. 1878 at Hitchin

♀ **Evelyn Jowitt**
b. 1880 at Stevenage

♀ **Audrey Jowitt**
b. 1881 at Hitchin

♂ **William Allen Jowitt**
b. 15 Apr 1885 at Stevenage
d. 1957, aged 72

previously advocated abolishing the university franchise, it is not surprising that he lost.

In 1936 he was re-admitted to the Labour Party and in October 1939 he was returned unopposed as MP for Ashton under Lyne. He was appointed Solicitor General in the coalition government by Winston Churchill in 1940 and in 1942 was appointed Paymaster General, in which role he had a remit to plan for post-war reconstruction. He continued this role in 1943 as a Minister without portfolio until 1944 when he was made Minister of Social Insurance. When the Labour Party won the 1945 election, his old school-friend Clement Attlee appointed him Lord Chancellor, at which point he was created Baron Jowitt of Stevenage. In January 1947 he was made Viscount Jowitt. While Lord Chancellor, he presided over the appeal of William Joyce (Lord Haw-Haw) against his conviction for treason. He remained Lord Chancellor until the Conservative victory of October 1951. He was then created Earl Jowitt, becoming leader of the opposition in the Lords.

For reasons which are not clear, Earl Jowitt pronounced his name to rhyme with 'know it' in contrast to his wool-dealing cousins who pronounced it to rhyme with 'how it'. This so infuriated J. H. Jowitt's daughter, Mary Caroline ('Moya') Jowitt, that she wrote him the following verses:

> My dear Cousin William, you're wrong and you know it
> When you like to be called the Lord Chancellor Jowitt.
> Although you don't come from the North Country now, it
> Is really correct to be Lord Chancellor Jowitt.
>
> Your family is Yorkshire, solid and Quaker,
> A fitting background for a famous law-maker.
> You ought to be proud that your forbears hit on
> The well-honoured trade of the wool that you sit on.[999]

Lord Jowitt wrote two books, the first of which was *The Strange Case of Alger Hiss* in 1953. It was criticised in right-wing American circles for what was perceived as its naïvety about the communist menace and by American jurists for displaying ignorance of the American judicial system. In 1954 he published *Some Were Spies,* touching on some of the cases he had tried as Solicitor General during the war. He also edited a *Dictionary of Law* which appeared after his death.

Although *The Times* described him as 'a brilliant advocate,'[1000] his reputation surely owes more to his astute political abilities. He died on 16 August 1957, aged seventy-two.

Notes

References to the Jowitt archives in the Brotherton Library, University of Leeds give both the Brotherton's own catalogue numbers and those given in Pat Hudson's *The West Riding Wool Textile Industry*. Sources marked (FTBJ) are in the possession of Mr F. T. B. Jowitt. See the Bibliography for full details of published works.

1.   Minutes, 26 July 1949.
2.   *Observations,* p. 37.
3.   Defoe 1724, vol. 3, p.98.
4.   Defoe 1724, vol. 3, pp. 114–15.
5.   Anon, *History of Leeds,* p. 44.
6.   Defoe 1724, vol. 3, p. 101.
7.   Anon, *History of Leeds,* p. 44. The author of this work, and another of other eighteenth-century sources, use the term 'mixed cloth' to mean what is less confusingly called 'mixture,' cloth made from several different colours of wool scribbled together before being spun. The book also mentions Archet as a place where white cloth was produced. There seems to have been no such place in the West Riding.
8.   Anon, *History of Leeds,* p. 23 - gives the wrong date, 1720, for its erection.
9.   Defoe 1724, vol.3, Letter I, pp. 116–7.
10. Defoe 1724, vol. 3, Letter I, pp. 119–21.
11. Defoe 1724, vol. 3, Letter I, pp. 107–8.
12. Originally a measure by volume, later 53 cwt, approximately 2.67 metric tonnes.
13. Defoe 1724, vol. 3, Letter I, p. 122.
14. Fraser, *History of Modern Leeds,* p. 120.
15. Dorian Gerhold, p. 496.
16. I must confess to some doubts about this. Although much reprinted, I have not managed to find the quotation in Leland's works
17. Riley, *Guide,* pp. 43–44.
18. Riley, *Guide,* p. 45.
19. Billam, *Walk,* 1806, p.2.
20. Riley, *Guide,* p. 34.
21. Riley, *Guide,* p. 36.
22. Riley, *Guide,* p. 36.
23. Billam, *Walks,* 1835, p. 109.
24. Billam, *Walks,* 1835, p. 70.
25. Billam, *Walks,* 1835, p. 118.
26. Meredith Baldwin Weddle, *Walking in the Way of Peace,* p. 44.
27. Anon, *An Account of the Constitution and Present State of Great Britain,* p. 65.
28. Housman, *Descriptive Tour,* p. 58.

29. Palmer, *A Fortnight's Ramble*, p. 32.

30. Radcliffe, *A Journey Made in the Summer of 1794*, pp. 385–6.

31. Nicholson, *Annals of Kendal*, p. 141.

32. Coleridge, Notebook, 1826, quoted in Gillman, *The Life of Samuel Taylor Coleridge*, pp. 246–7.

33. Cooley, *A Cyclopedia of Practical Receipts*, p. 252, which also gives the following recipe:
   Take ½ lb. of opium, sliced; 3 pints of good verjuice; 1½ oz. of saffron; boil them to a proper thickness, then add ¼ lb. of sugar and two spoonfuls of yeast. Set the whole in a warm place, near the fire, for 6 or 8 weeks, then place in the open air until it becomes of the consistence of sirup; lastly, decant, filter, and bottle it up, adding a little sugar to each bottle.

34. Robert Jowitt, Diary 1846–7, entry for 27 March 1847.

35. Besse, *Sufferings*, vol. 2, p. 104.

36. Wails, *Warning*, p.10.

37. Thoresby, *Diary*, p. 168.

38. Thoresby, *Ducatus Leodiensis*, pp. 211–12.

39. Thoresby, *Ducatus Leodiensis*, p. 219.

40. Probate documents, Borthwick Institute, York.

41. Besse, *Sufferings*, vol. 2, pp. 154–158.

42. That Hannah was Richard's daughter is clear from the Quaker minute books. The evidence for the young Tabitha (not to be confused with Richard's widow, Tabitha, *née* Hopwood) is more circumstantial. She was in Beeston at the right time and she appears on earlier family trees. However, the dates given for her are those of Hannah Jowitt.

43. *Leeds Friends' Minute Book: 1692–1712*.

44. Probate documents, Borthwick Institute, York.

45. Mortimer, *Leeds Friends' Minute Book*, Biographical Notes, p.213.

46. Minutes of Leeds Preparative Meeting, 26 June 1697. Brotherton E1, p. 31.

47. Minutes of Leeds Preparative Meeting, 26 June 1697. Brotherton E1, p. 70.

48. Minutes of Leeds Preparative Meeting, 26 June 1697. Brotherton E1, p. 31.

49. *Leeds Friends' Minute Book: 1692–1712*, p.77.

50. *Leeds Friends' Minute Book: 1692–1712*, pp. 92, 93.

51. There were many Gotts in the West Riding at the time and there is no indication that this John is closely related to the famous Benjamin Gott of Leeds.

52. Minutes of Leeds Preparative Meeting, 26 June 1697. Brotherton E1, p. 74.

53. Minutes of Leeds Preparative Meeting, 14 December 1720. Brotherton E2, p. 60.

54. Hirst, *History of the Woollen Trade*, p. 44.

55. Coleridge, *Letters, pp.* 121 (n), 157.

56. *Wesleyan-Methodist Magazine for 1822*, pp. 396–6.

57. See Samuel Birchall, *Descriptive List*.

58. *London Evening Post* 24 January 1755.

59. Brotherton 30, Hudson 47.

60. Robert Benson Jowitt in John Jowitt, *Reminiscences*, p. 5.

61. John Jowitt Jun. [i.e. John Jowitt II] Ledger (Botherton 38, Hudson 1).
62. Hudson, *Genesis of Industrial Capital,* p. 122.
63. Hopper, 'The Farnley Wood Plot'.
64. Brotherton, Manuscripts: MS Dep. 1979/1 (Carlton Hill Archive) Gildersome School.
65. http://homepage.eircom.net/~lawedd/TOOTHILL.htm
66. Bischoff, vol. 1, pp. 146–7.
67. Brotheron 30, Hudson 47. A similar notice was placed by wool-buyers in the *Bath Chronicle,* 4 June 1789.
68. John Jowitt Jun. [i.e. John Jowitt II] Ledger (Botherton 38, Hudson 1). Nevins & Gatliff invested heavily in steam-powered machinery supplied by Boulton & Watt and this is reflected in their insurance valuation of £10,400 in about 1795 – Chapman, 'Fixed capital formation,' pp. 259–60.
69. Somervell, *Isaac and Rachel Wilson,* p 24, etc.
70. Silberling.
71. Ward, *The factory movement,* p. 9.
72. Hirst, *History of the Woollen Trade,* pp. 10–11.
73. Hirst, *History of the Woollen Trade,* pp. 9, 10.
74. Letter from John, Joseph and Samuel to John Syder & Son, Windham [Wymondham, Norfolk?], 17 April 1802.
75. Advertisement in the *Leeds Mercury,* Saturday, 8 April 1815
76. Robert Jowitt Private Ledger (Brotherton 2, Hudson 27).
77. John Jowitt Jun. [i.e. John Jowitt II] Ledger (Botherton 38, Hudson 1), pp. 95, 115). John Jowitt II had already lent Samuel Birchall £500 on 10 February 1793.
78. *Leeds Mercury,* 18 September 1824.
79. *London Gazette,* 7 July 1821, p. 11.
80. *London Gazette,* 2 January 1838, p. 11.
81. Tithe maps, 1836–51.
82. Robert Jowitt, Diary 1846–7, entry for 8 February 1847.
83. Sometimes spelt Rachel by members of the family.
84. Somervell, *Isaac and Rachel Wilson,* p. 47.
85. Somervell, *Isaac and Rachel Wilson,* pp. 51–67.
86. Rachel Wilson spoke out strongly against it – Somervell, *Isaac and Rachel Wilson,* p. 51.
87. Hudson, *Genesis of Industrial Capital,* p. 123.
88. Bischoff, vol. 1, p. 408.
89. Letter from John, Joseph and Samuel to John Syder & Son, Windham [Wymondham, Norfolk?], 17 April 1802.
90. Letter to Mr Elliston, 3 October 1805 (Brotherton 31, Hudson 48).
91. Letter to Joseph Keep, Colchester, 3 June 1802 (Brotherton 31, Hudson 48).
92. Letter to William Oakley, 3 October 1805 (Brotherton 31, Hudson 48).
93. Letters to: Captain Hoile, January 1803; George Finch & Co. 28 January 1803 (Brotherton 31, Hudson 48).
94. Letter to Elsted Warnham, 2 February 1804 (Brotherton 31, Hudson 48).
95. Robert Jowitt and John Jowitt junior, diaries for 1831–2.

96. Probate records.
97. O'Donoghue.
98. Robert Jowitt, quoted in *Quarterly Journal of Agriculture*, vol. I, May 1828–August 1829, p. 376.
99. 6d per lb according to Pat Hudson, *Limits of Wool*.
100. House of Commons, Minutes of Evidence, 1833.
101. John Jowitt jr, Private Ledger A (Brotherton 23, Hudson 37). These figures agree very closely with those drawn up in 1900 by RBJ (Appendix IV).
102. Hudson, 'Genesis of Industrial Capital', p. 123.
103. Robert Jowitt Private Ledger (Brotherton 2, Hudson 27).
104. His father, John II, owned a single share in the Leeds waterworks worth £120 – John Jowitt II ledger (Brotherton 38, Hudson 1)
105. Odlyzko, *This Time is Different.*
106. Brotherton, Robert Jowitt Private Ledger (Brotherton 2, Hudson 27).
107. This refers to the annual Woodhouse Feast which started on the Sunday on or after 21 September.
108. 'Original Essay on Temperance,' *Leeds Mercury*, 7 December 1833.
109. Cashbook and Ledger C 1854–62 (Brotherton 16, Hudson 29).
110. Brotherton, Robert Jowitt Private Ledger (Brotherton 1, Hudson 28).
111. *Annual Monitor for 1864,* pp. 45–8.
112. Robert Jowitt, Diary, 16 March 1832.
113. From which he resigned in 1825 – *Leeds Mercury*, 9 July 1825.
114. *Friends for Life,* p. 183.
115. *Leeds Mercury*, 16 February 1858.
116. Robert Jowitt, Diary for 1822–1825, entries for 9 December 1822, 28 January 1823.
117. John junior, Diary for 1835–1836, 21 May 1835.
118. Robert Jowitt, Diary.
119. Robert Jowitt, Diary 1822–5, entry for 9 September 1824.
120. Robert Jowitt, Diary 1832, entry for 26 April 1832.
121. Diary for 1831–2, transferred from temporary notebook covering roughly the period between 20 October and 10 November 1831.
122. It is sometimes difficult to distinguish in newspaper reports between the two John Jowitts, except that Joseph's son lived in Hanover Square and is usually referred to as John Jowitt Esquire. John Junior is only referred to as Esquire very late in life.
123. This is presumably Thomas Mercer, whose Classical and Commercial Academy was conveniently located near Robert Jowitt's office in Albion Street – *Leeds Mercury*, 3 January 1818. His *Vocabulary, Latin and English, on a New Plan* was published by Longman, Hurst, Rees, Orme & Browne in 1817.
124. Robert Jowitt, Diary, 6 March 1824.
125. Robert Jowitt, Diary, 12 July 1824.
126. John Jowitt junior, Diary for 1831–2, undated entry.
127. Robert Benson Jowitt in *Reminiscences,* p. 7.
128. John Jowitt junior, Diary 1835–6.

129. Robert Benson Jowitt in *Reminiscences*, p. 7.
130. *Oxford Dictionary of National Biography*.
131. Susan Maria Howard, in *Reminiscences*, p. 13.
132. Anna Dora in John Jowitt, *Reminiscences*, p. 22.
133. John Jowitt junior, Diary 1831–2, undated entry but November 1831
134. John Jowitt, *Reminiscences*, p. 17.
135. John Jowitt jr, Private Ledge A (Brotherton 23, Hudson 37).
136. John Jowitt junior, Diary for 1831–2, 23 September 1831. The locomotive re-ferred to by John junior may have been the very first built by Galloway, Bow-man and Glasgow, the *Manchester*.
137. John Jowitt jr, Private Ledge A (Brotherton 23, Hudson 37)
138. John Jowitt jr, Private Ledge A (Brotherton 23, Hudson 37)
139. Robert Benson Jowitt in *Reminiscences*, p. 8.
140. John junior, Diary, 26 September 1831.
141. Elizabeth married the banker Alfred Harris (1801–1880) of Oxton Hall near Tadcaster on 2 April 1835.
142. John Jowitt junior, Diary for 1831–2, 19 October.
143. Hagen (1805–1870) emigrated to Australia in 1839 where he proved himself to be a ruthless businessman.
144. John Jowitt junior, Diary for 1831–2. The exact date of this entry is unclear.
145. John Jowitt junior, Diary for 1831–2
146. John Jowitt junior, Diary for 1831–2, transferred from temporary notebook. The exact date is unclear – between about 20 October and 10 November 1831.
147. 'I sometimes fear she is not happy in the prospect, because so desirous to retard it.' John Jowitt junior, Diary for 1835–6, 3 December 1835.
148. John Jowitt junior, Diary for 1831–2, 22 September 1831.
149. John Jowitt junior, Diary for 1835–6, 10 April 1835.
150. John Jowitt junior, Diary for 1835–6, 5 October 1835.
151. John Jowitt junior, Diary for 1835–6.
152. John Jowitt junior, Diary for 1835–6.
153. John Jowitt junior, Diary for 1835–6.
154. John Jowitt junior, Diary for 1835–6, entry for 7 September 1835.
155. 'Cousin W,' probably a Wilson.
156. John junior seems to have translated the German *der See* [lake] as though it were *die See* [sea].
157. Wordsworth, *Memorials of a Tour*, pp. 60, 64.
158. Daughter of Joseph Gurney and wife of Jonathan Backhouse, a leading Quak-er banker who helped to finance the Stockton & Darlington Railway. He was also present at the wedding.
159. Robert Jowitt, Diary, 5 May 1836. I have only seen an extract of this diary and do not know where it is.
160. Moses Brown in a letter to John Wilbur, 4 July 1836, in Wilbur, *Journal*, p. 177.
161. This is presumably the church which stood there before the present St Pat-ricks was built in the 1850s. The pedantic sermon was almost certainly deliv-ered by the Rev. John Thompson.

162.   Robert Benson Jowitt in *Reminiscences,* pp. 7–8.

163.   Deborah Jowitt, Diary, 20 May 1836.

164.   John Jowitt junior, Diary for 1835–6, entry for 29 September 1835.

165.   Crewdson, *Beacon,* p. 5.

166.   On 12 January 1835 for two shillings, John Jowitt jr Private ledger A (Brotherton 23, Hudson 37).

167.   Stephen Gould of Newport, Rhode Island, to John Wilber, 24 December 1836, in Wilbur, *Journal,* p. 214.

168.   John Jowitt junior, Diary for 1835–6, entry for 19 April 1835. John junior did not attend the meeting but relied on a report from his uncle, Isaac Wilson.

169.   John Jowitt junior, Diary for 1835–6, entry for 18 April 1835.

170.   John Jowitt junior, Diary for 1835–6, entry for 21 November 1835. It is hard to see what the relevance of mothers-in-law and daughters-in-law is.

171.   Robert Jowitt, *Thoughts on Water Baptism,* p. 24.

172.   John Jowitt junior, Diary for 1831–2, entry for 6 December 1831.

173.   Wake, *Kleinwort Benson,* p. 52.

174.   Society of Friends, *Inquirer,* November 1838.

175.   *Friend of India* (Calcutta), 18 April 1839.

176.   *Journal of the Friends' Historical Society,* vol. 53, 1975, p. 62,

177.   Excessive male neonatal mortality sometimes indicates a genetic defect on the X chromosome and the death of these five sons, all born between Anna Dora in 1844 and Emily in 1853, may be a consequence of intermarriage in the Quaker community.

178.   John Jowitt junior, *Reminiscences.*

179.   John Jowitt junior, *Reminiscences,* p. 16.

180.   Letter to Susan Maria, c. 1884, John Jowitt junior, *Reminiscences,* pp. 106–7.

181.   1851 census.

182.   John Dearman Birchall, *Victorian Squire,* p. xii.

183.   Tithe map.

184.   *Yorkshire Herald & York Herald,* 6 September 1900.

185.   John Jowitt Jr, Private Ledger B (Brotherton 24, Hudson 38).

186.   *Leeds Mercury,* 8 September 1865. It may be at this point that RBJ, who had been brought up a Baptist, converted to the Church of England.

187.   John Dearman Birchall, *Victorian Squire,* p. 5.

188.   *Medical Directory for Scotland,* 1854, pp. 188–9, which also said, 'The memory of so good a man will dwell in the minds of young and old, who should imitate his example, and walk in his footsteps.'

189.   Thom, *Sketch.*

190.   Letter to Robert Benson Jowitt, 22 May 1852, in John Jowitt, *Reminiscences,* p. 39

191.   John Dearman Birchall, *Victorian Squire,* p. 19.

192.   Probate record. The full inventory is given in Appendix I.

193.   J. D. Birchall & Co. owned the Burley Mill in Kirkstall which had been built by the great pioneer of mechanisation, Benjamin Gott, in 1798. Their cloths (some bought in rather than manufactured by themselves), including wool-

lens, worsteds and tweeds, were widely exhibited abroad and won many prizes. The company, a partnership between Dearman, Oswald Birchall, Archibald Campbell and Emil Kafka, was dissolved on 31 August 1891 – *London Gazette,* 4 September 1891.

194. Robert Benson Jowitt in John Jowitt, *Reminiscences,* p. 10.
195. William Forster, *Memoirs of William Forster,* vol. 1, p. 42.
196. R. J. & S. Private Business Letters (Brotherton 37, Hudson 54).
197. Obituary in Society of Friends, *Extracts from the Meetings and Proceedings,* Appendix D, pp. xii–xv.
198. R. Jowitt, Letter Book 1853–6.
199. John Jowitt junior, Diary, 31 December 1836.
200. RBJ notes (FTBJ).
201. Crewdson's diary (now lost), quoted in Mrs Dorothy Jowitt's unpublished account of Robt. Jowitt & Sons.
202. L Ledger B (Brotherton 44, Hudson 13), p. 266.
203. Death certificate.
204. L Ledger B (Brotherton 44, Hudson 13), p. 266.
205. R. B. Jowitt in John Jowitt, *Reminiscences,* p. 6.
206. Brotherton, John Jowitt jr, Private Ledger A (Brotherton 23, Hudson 37).
207. Brotherton, John Jowitt jr, Private Ledger A (Brotherton 23, Hudson 37).
208. Brotherton, Account of Failures, (Brotherton 6, Hudson 62).
209. R. O. Roberts, 'Bank of England Branch Discounting,' p. 231.
210. John Jowitt & Co. was the wool staplers run by Joseph's son, John senior, from premises in Westgate.
211. *Solicitors' Journal and Reporter,* 13 April 1861, p. 425.
212. *Morning Post,* 9 November. 1860.
213. Beresford, *Leeds Chamber of Commerce,* p. 147.
214. *The Times,* 22 April 1861.
215. Robert Benson Jowitt in John Jowitt, *Reminiscences,* p. 9.
216. *Leeds Mercury,* 28 August 1858; 15 February 1859; 17 January 1860.
217. *Leeds Mercury,* 5 December 1846.
218. Robert Benson Jowitt, in John Jowitt, *Reminiscences,* pp. 8, 9.
219. *Leeds Mercury,* 16 December 1858.
220. *Leeds Mercury,* 6 February,1858.
221. *Leeds Mercury,* 29 November 1851.
222. *Leeds Mercury,* 6 February,1858. The Rev. Hook, vicar of Leeds, was high-church Anglican but strangely at home with non-conformists such as John Jowitt junior and Edward Baines.
223. Robert Benson Jowitt in John Jowitt, *Reminiscences,* pp. 9, 10.
224. *Leeds Mercury,* 8 November 1862.
225. *Leeds Mercury,* 25 November
226. Letter dated 12 December, *Leeds Mercury,* 13 December 1862.
227. John Dearman Birchall, *Victorian Squire,* p. 50.
228. Robert Benson Jowitt's diary, now missing, quoted in Mrs Dorothy Jowitt's unpublished account of Robt. Jowitt & Sons.

229. Or December 1883, according to Dorothy Jowitt.
230. John Jowitt, *Reminiscences,* various pages.
231. Christmas 1852?, on RBJ's coming home from college in John Jowitt, *Reminiscences,* p. 43.
232. *Daily News,* 4 July 1856.
233. Barnard, 'Wool Buying in the Nineteenth Century,' p. 2.
234. Anon, *Leeds and Bradford.*
235. Barnard, 'Wool Buying in the Nineteenth Century,' p.3.
236. Barnard, 'Wool Buying in the Nineteenth Century,' p.8.
237. Huurdeman, p.123.
238. Robt. Jowitt & Sons Private Business Letters (Brotherton 37, Hudson 54), unnumbered.
239. Robt. Jowitt & Sons Private Business Letters (Brotherton 37, Hudson 54), unnumbered.
240. Robt. Jowitt & Sons Private Business Letters (Brotherton 37, Hudson 54), unnumbered.
241. Brotherton 51, Hudson 63.
242. Robert Benson Jowitt Private Ledger B (Brotherton 25, Hudson 41). While John junior does seem to have been tea-total, Benson's grandfather, Robert, who supported the temperance movement, does not seem to have been. His ledger (Brotherton 16, Hudson 29, p. 117) shows modest purchases of port and sherry, including 2 dozen bottles of amber sherry in 1858.
243. His home in Tunbridge Wells, Hurstwood Lodge, was described as a beautiful house in an acre of grounds with 'a remarkable rock garden' – *The Times,* 28 June 1921.
244. Teale, *Historical Sketch.*

245. Teale, *Historical Sketch,* pp. 27–8.
246. Letters, *The Times,* 11 January 1909.
247. *Leeds Mercury,* 25 September 1886.
248. *Leeds Mercury,* 3 November 1860.
249. *Leeds and West Riding Express,* quoted in the *The Musical World,* 24 November 1860, p. 745. The Micromegas reference is to a short story by Voltaire.
250. *Leeds Mercury,* 29 January 1885.
251. Many members of the Benson family were called Robert. In this book, Robert Benson refers specifically to John junior's father-in-law, the inhabitant of Parkside, Kendal.
252. *Niles Weekly Register,* vol. 10, 1816, p. 321.
253. *Gentleman's Magazine,* vol. 38, 1835, p. 204. Blythe is also thought to have studied chemistry in Manchester under John Dalton. He filed several patents relating to textile dyes with John Mercer, the inventor of mercerisation.
254. *Preston Guardian,* 25 August 1849.
255. *Manchester Times,* 18 September 1850.
256. M. E. Kopf.
257. Advertisement for a 'Fancy Dyer... well acquainted with garancine work,'

Manchester Guardian, 7 August 1847.

258. Source: F. T. B. Jowitt.
259. G. F. Benson, *Historical Record,* p. 18.
260. Letter from John junior to W. T. Benson, 12 March 1870, enclosing duplicate invoice for nine bales of wool, Private Copying Book (Brotherton 36, Hudson 53); note on delivery instructions (to be shipped to Boston rather than Montreal) and interest rates, Jobbing Orders (Brotherton 12, Hudson 61), pp. 16–17. The two companies were still doing business together in 1920 – Customers ledger (Brotherton 70, Hudson 80), pp. 563–5.
261. Robt. Jowitt & Sons Private Business Letters (Brotherton 37, Hudson 54).
262. Robt. Jowitt & Sons Private Business Letters (Brotherton 37, Hudson 54), fol. 75.
263. Robt. Jowitt & Sons Private Business Letters (Brotherton 37, Hudson 54), fol. 81.
264. Robt. Jowitt & Sons Private Business Letters (Brotherton 37, Hudson 54).
265. Salzedo and Brunner, *Briefcase on Contract Law,* pp. 50–51.
266. *London Gazette,* 19 August 1890.
267. Poulter, *Early History of Electricity Supply.*
268. Robt. Jowitt & Sons Private Business Letters (Brotherton 37, Hudson 54), p. 61.
269. *The County Gentleman: Sporting Gazette, Agricultural Journal, and The Man about Town,* 25 July 1891
270. Advertisement in *Leeds Mercury,* 22 January 1878.
271. Referred to in the Memoranda (Brotherton 25, Hudson 41) as simply J. F. W., but the very fact that he is referred to by initials alone suggests a member of the extended Jowitt family or close acquaintance, and his marriage to George Portway's sister confirms the identification. George and Charlotte Portway were born in Bury St Edmunds but the family, headed by George senior, had settled at Springfield Mount, Leeds.
272. Wilson's grandparents were Isaac and Mary Wilson (née Jowitt). Mary was one of John II's daughters.
273. Robert Benson Jowitt's diary, now missing, quoted in Mrs Dorothy Jowitt's unpublished account of Robt. Jowitt & Sons.
274. Dorothy Jowitt's unpublished account of Robt. Jowitt & Sons.
275. Robert Benson Jowitt, Private Ledger B, (Brotherton 25, Hudson 41), Memoranda.
276. William Musgrave Wood (1842–1914) was the son of Joseph Wood and Ann Musgrave. He inherited Swinnow Mill from his father and ran it for a while as W. M. Wood & Co. The 1836–51 tithe maps show that Joseph Wood owned a total of about 13 acres of land, including the mill, fields, a house and cottages.
277. *London Gazette,* 2 January 1891.
278. Robert Benson Jowitt, Private Ledger B, (Brotherton 25, Hudson 41), Memoranda.
279. *London Gazette,* 2 March 1897, p. 1286.
280. Extracted from the correspondence files (Brotherton 37, Hudson 54).

281. According to a note to the Surveyor of Taxes, Robt. Jowitt & Sons Private Business Letters, 4 July 1904 (Brotherton 37, Hudson 54), p. 300.
282. Letter, 1900, Robt. Jowitt & Sons Private Business Letters, p. 220
283. RBJ notes (FTBJ).
284. The 1881 census also shows that Frederick attended the preparatory school at Hartley Wintney, Hampshire.
285. 'Of the individual players, Jowitt, for Hertford, showed decidedly the best form; he keeps goal in excellent style, and never loses his head.' – *Oxford Magazine*, vol. 4, 1887, p. 414; he played cricket for the Suffolk Borderers in their matches against Ipswich School and Harleston District, a poor batsman but adequate fielder and bowler – *Ipswich Journal*, June 25, 1885 and *Bury and Norwich Post and Suffolk Herald*, July 14, 1885; and he ran modestly well in the 100 and 150 yard handicaps in the Hertford College sports day – *Bell's Life in London*, 18 February 1886.
286. Jobbing Orders (Brotherton 12, Hudson 61), p. 109.
287. Jobbing Orders (Brotherton 12, Hudson 61), p. 82.
288. *Hawkes Bay Herald*, 8 March 1889.
289. *Otago Witness*, 15 February 1894; 16 November 1893.
290. Samuel Hales of Lismore Street, Lawrence, Otago, died in 1884.
291. 'Alice's Letter to Her Readers,' *Otago Witness*, 25 February 1892.
292. He represented the Otago Golf Club at a match in Christchurch – *Otago Daily Times*, 20 July 1894
293. He was one of the lowest-scoring batsmen in the Carisbrook Cricket Club B team – *Otago Witness*, 4 June 1896
294. His election is recorded in the *Otago Daily Times*, 21 February 1893 and his resignation in the *Otago Witness*, 2 September 1897.
295. *Otago Witness*, 15 April 1897
296. *Otago Witness*, 8 July 1897.
297. Advertisement, *Otago Witness*, 29 April 1897.
298. *New Zealand Graphic and Ladies' Journal*, Saturday, 22 May 1897.
299. RBJ, Diary, 25 February 1897, quoted by David Jowitt (private correspondence).
300. RBJ, Diary, 3 August, quoted by David Jowitt (personal communication).
301. RBJ, Diary, 4 August, quoted by David Jowitt (personal communication).
302. RBJ, Diary, 4, 7, 11 and 12 August 1897, quoted by David Jowitt (personal communication).
303. RBJ, Diary for 16–26 August, quoted by David Jowitt (personal communication).
304. Letter from Mrs Hales,in RBJ, Diary, 28 August 1897, quoted by David Jowitt (personal communication).
305. Letter from Frederick McC. Jowitt to Rinah undated, quoted by David Jowitt (personal communication).
306. I am grateful to Sabrina Zinke, Deputy Archive Manager at the University, for this information.
307. I am grateful to Robert Ernald Jowitt for this information. The Simpsons

came from Norwich, where Reginald's father, George, practised as a solicitor.

308. Source R. E. Jowitt.
309. *The Times,* 5 January 1895.
310. Sheffield, *Works,* vol. 1, p. 199.
311. *London Gazette,* 9 February 1906 indicates that the partnership, consisting of R. B. Jowitt, F. McC. Jowitt and Robert Jowitt would continue the business of wool merchants, brokers and top makers, while William Gibson would continue the fellmongery. RBJ's diary for 1899 indicates that the hc received a letter from the Vicar of Meanwood complaining about the stench from the plant (then managed by RBJ's son Robert) – quoted by Dorothy Jowitt, unpublished account.
312. G. F. Benson, *Historical Record,* p. 95.
313. Brotherton, Papers of the Lancashire Cotton Districts Relief Fund, 40.
314. Dorothy Jowitt, unpublished account.
315. He visited his mother-in-law in Canada 1907 and 1908.
316. Victoria University of Manchester, *Register of Graduates.* Alfred Roland Hummel graduated in 1905 from the University of Leeds, until 1904 a constituent of the Victoria University. Alfred's brother Raymund also worked for the company.
317. Robt. Jowitt & Sons Private Business Letters (Brotherton 37, Hudson 54), P/L summary pp. 407–9.
318. All these companies had London offices – Trowbridge Sample Book (Brotherton 110, Hudson 143).
319. Peter Bell, *Robert Jowitt & Sons, the Limited Company.*
320. Letter to R. B. McComas, 28 June 1906, Robt. Jowitt & Sons Private Business Letters (Brotherton 37, Hudson 54), unnumbered.
321. Robt. Jowitt & Sons Private Business Letters (Brotherton 37, Hudson 54), fol. 208.
322. Robt. Jowitt & Sons Private Business Letters (Brotherton 37, Hudson 54), unnumbered.
323. Robt. Jowitt & Sons Private Business Letters (Brotherton 37, Hudson 54), fol. 314.
324. Robt. Jowitt & Sons Private Business Letters (Brotherton 37, Hudson 54), unnumbered.
325. Letter from Frederick to Victor, 4 January 1907, Robt. Jowitt & Sons Private Business Letters (Brotherton 37, Hudson 54), unnumbered.
326. Letter from Frederick to Sheard, 9 July 1907, Robt. Jowitt & Sons Private Business Letters (Brotherton 37, Hudson 54), unnumbered.
327. Robt. Jowitt & Sons Private Business Letters (Brotherton 37, Hudson 54), unnumbered.
328. 14 Feb. 1908, Robt. Jowitt & Sons Private Business Letters (Brotherton 37, Hudson 54), unnumbered.
329. 16 April 1908, Robt. Jowitt & Sons Private Business Letters (Brotherton 37, Hudson 54), unnumbered.
330. Source: T. P. Sykes.

331. Letters to McComas, 28 Jun 1906 and to Edwards, 5 October 1906, Robt. Jowitt & Sons Private Business Letters (Brotherton 37, Hudson 54), unnumbered.

332. 5 October 1906, Robt. Jowitt & Sons Private Business Letters (Brotherton 37, Hudson 54), unnumbered.

333. *Brisbane Courier,* 28 May 1896.

334. *London Gazette,* 24 June 1902.

335. Produce ledger, January 1907–July 1908 (Brotherton 67, Hudson 26).

336. 18 April 1908, Robt. Jowitt & Sons Private Business Letters (Brotherton 37, Hudson 54), unnumbered.

337. 18 April 1908, Robt. Jowitt & Sons Private Business Letters (Brotherton 37, Hudson 54), unnumbered.

338. Minutes, 8 April 1932.

339. *Financial Times,* 19 April 1913.

340. *London Gazette,* 5 July 1912.

341. *The Times,* 21 April 1913.

342. Sources: Peter Varey; probate records.

343. Branch Ledger K (Brotherton 73, Hudson 72), pp. 11–40, 115, 118, 120. See also Appendices VI and VII.

344. Letter, 5 August 1914 (FTBJ).

345. Letters from F. McC. J., 11 & 14 August 1914 (FTBJ).

346. For instance, £300 to the Lord Mayor's Distress Fund, 23 May 1917 and donations to the RNIB, orphanages, hospitals, the Flower Girls' Mission and for the relief of Jews in Palestine – Ledger K (brotherton 73, Hudson 72), pp. 216, 217, 235.

347. Source: FTBJ.

348. Frederick Jowitt and A. T. Keeling had met at Marlborough and Frederick watched Queen Victoria's Golden Jubilee procession from the Keelings' balcony in Cockspur Street in 1887 – Dorothy Jowitt, unpublished history of Robt. Jowitt & Sons, p.21.

349. A. T. Keeling, Memorandum. See Appendix IX.

350. By 1924, Barclays were providing an overdraft facility of £250,000. In 1927, the directors were required to give personal guarantees against £50, 000 of the overdraft in the following proportions: E. M. Jowitt 22%, S. Harland 22%; R. Jowitt, 16%, W. T. B. Jowitt and R. B. Jowitt 15% each, G. Blackwell 10%. These guarantees remained until June 1940. Minutes, 18 March 1927 & 5 June 1940.

351. Wool and materials £25,000; machinery £5,000; consequential loss £20,000.

352. Minutes, 23 January 1920.

353. http://www.leodis.org/.

354. *London Gazette,* 9 February 1906.

355. Minutes, 19 December 1919.

356. Minutes, 10 January 1922.

357. *London Gazette,* 30 June 1922.

358. Minutes, 15 June 1922.

359. *London Gazette*, 20 June 1922.

360. 1 bale of 64's quality greasy Australian Merino fleece wool was equivalent to about 0.73 pack of top in the 1970s. Bale weights were lower in the pre-war years. Cape wools had lower yields than equivalent Australian wool types. Tear for 64's wool was about 12:1 ration of top:noil. Source: Peter Bell.

361. Minutes, 2 June 1920.

362. Anning, *The General Infirmary at Leeds: the Second Hundred Years,* vol. II, p. 89.

363. Obituary, *British Medical Journal,* 24 August 1963, p. 506.

364. Notice of ninetieth birthday, *British Medical Journal,* 14 September 1957, p. 652.

365. Paper delivered to Association of Registered Medical Women, 5 May 1908 – *British Medical Journal,* 16 May 1908, p. 1178.

366. Paper delivered to Association of Registered Medical Women, 3 January 1911 – *British Medical Journal,* 21 January 1911, p. 139.

367. Minute books.

368. Minute, 4 November 1921.

369. Part of the first 200,000 shares issued, in exchange for the proportion of the partnership held by the trust set up to pay Caroline's annuity.

370. Minutes.

371. Peter Bell, *Robert Jowitt & Sons, the Limited Company.*

372. Reprinted in *Essays in Persuasion.*

373. *Australasian,* 9 May 1896 & *Argus* (Melbourne), 4 May 1896.

374. *Sunday Times* (Perth), 16 April 1933.

375. Minutes, 25 July 1951; letter from Frederick McC. Jowitt to H. B. McComas, 28 June 1906. Frederick wrote,

     As you know, we intend to start a Sydney Branch this year, and as I promised my late partner, Mr. Wm. Gibson, to give his nephew a chance, we are giving him the local managership. As you knew and appreciated the sterling qualities of John Gibson's father, our late manager in Melbourne, may I ask you, if you happen to be in Sydney, to do what you can for young John Gibson. He is a decent boy, and means well I think, but so far he doesn't show any great driving power, and unless he puts his back very much into the work, and keeps his eyes open, he will not do much good, I fear, in Sydney.

376. John had his previous debts to the company written off and was warned in a letter from Frederick McC. Jowitt on 4 January 1907, 'Of course, in the future you must keep within your salary and must not get over-drawn.'

377. Minutes, 17 March & 24 April 1925.

378. Minutes, 22 July 1927.

379. *Western Mail* (Perth), 30 January 1930; *Advertiser* (South Australia), 27 January 1930; *Register News-Pictorial,* 27 January 1930.

380. Minutes, 31 January 1930.

381. G. A. Reid joined Jowitts from the rival company, Richard Moore & Son of East London in 1907. Frederick McC. Jowitt left the hapless Reid awaiting a final confirmation for several months while it placated Moores. The company

wrote to Reid on 7 June 1907,

> We have received from you the following cables: 31st May, 'Yes'; June 3rd, 'Shall I resign this mail?'; June 5th, 'Cable from Moore replies accepting Jowitt's offer finally'. From these we conclude that you have finally decided to accept our offer and that all is in order.

Frederick went on to explain that, 'it would be suicidal for Moores and ourselves, out of ill-feeling or jealousy, to run each other up, and give large prices to store-keepers.'

382. Minutes, 27 May 1928.
383. Minutes, 27 July 1938.
384. Minutes, 19 December 1919.
385. Minutes, 23 January 1920.
386. Minutes, 10 January 1922.
387. Robt. Jowitt & Sons Private Business Letters (Brotherton 37, Hudson 54), p. 236, letter to Slaughter & May, 25 May 1901; p. 237, letter to Thomas Jackson & Co., 29 May 1901.
388. Minutes, 26 March 1920.
389. Minutes, 23 April 1920.
390. Minutes, 11 March 1921.
391. Minutes, 11 March 1921.
392. Minutes, 15 June 1921.
393. Memorandum of proposed arrangement, included with minutes, 23 September 1921.
394. Minutes, 23 September 1921.
395. Letter from Magnus Irvine to A. Leonard Scott, 8 August 1921, included with Minutes.
396. Minutes, 4 November 1921.
397. Minutes, 10 January 1922.
398. They were sealed with the RJ&S seal on 21 July 1922 (Minutes for that date).
399. Minutes, 11 May 1922.
400. Extract from minutes of Walter Scott & Sons Ltd, 28 July 1922, filed with RJ&S minutes.
401. *Scotsman*, 29 July 1922.
402. Minutes, 23 February 1923.
403. Minutes, 23 February 1923.
404. Minutes, 9 April 1923.
405. Minutes, 13 April 1923.
406. 'By supplying some dyed yarn into the drafting zone, it is also possible to produce a "splash yarn"... The splash yarn device produces a yarn that has scattered colour.' – Hergeth.
407. Minutes of joint RJ&S and WS&S board meeting, 4 May 1923.
408. Minutes, 28 September 1923.
409. Minutes, 11 January 1924.
410. Minutes, 29 January 1924.
411. The possibility of buying the property was discussed by the Board (minutes,

18 July 1924) and was clearly acted upon.

412. Minutes, 18 March 1927.
413. Minutes, 23 October 1931.
414. Minutes, 30 November 1925.
415. Minutes, 22 July 1927.
416. Minutes, 22 May 1924. Prince Sergei Obolensky was married to Tsar Nicholas II's daughter, Catherine Alexandrovna Yurievskaya, until they divorced in 1924. Having served during both the First World War and the Russian Civil War, he served as a Lieutenant Colonel in the U. S. paratroopers during the Second World War. He was also a member of the Office of Strategic Services and Vice-Chairman of Hilton Hotels.
417. Minutes, 24 October 1924.
418. Minutes, 7 February 1925.
419. Minutes, 3 May 1929.
420. Leonard, *Secret Soldiers,* p. 75.
421. National Archives, KV 3/35 document summary.
422. *The Times,* 20 May 1927.
423. *Hansard.*
424. Cabinet Minutes, National Archives, CAB/23/55.
425. *The Times,* 30 April 1929.
426. National Archives, KV 2/2559.
427. National Archives, KV 2/2684.
428. *Northern Territory Times,* 15 July 1927.
429. Minutes, 19 July 19929.
430. *Courier–Mail* (Brisbane), 10 February 1934.
431. http://codexsinaiticus.org/en/codex/history.aspx
432. Minutes, 12 June 1931.
433. Minutes, 23 October 1931.
434. Minutes, 29 September 1933.
435. Minutes, 20 April & 8 June 1934.
436. Churchill consulted widely. Unfortunately, he followed the advice of Montagu Norman and Sir Otto Niemeyer, Financial Controller at the Treasury, rather than the opposing advice of Keynes and John Bradbury. Churchill later described his decision on the Gold Standard as the worst mistake he ever made.
437. *The Times,* 15 April, 17 April, 17 May, 27 May, 4 June 1930.
438. Minutes, 5 July 1930.
439. Minutes and Peter Bell, *Robert Jowitt & Sons, the Limited Company.*
440. *Otago Witness,* 7 November 1900. One of the wool merchants who suspended payments during this crisis was Jules Florin. This must have been merely a temporary embarrassment as Jowitts were employing Jules and André Florin as agents in Roubaix in the thirties.
441. Minutes, 28 December 1931.
442. Minutes, 28 December 1931.
443. Minutes, 19 February 1932.

444. Minutes, 28 May 1932.
445. Minutes, 1 July 1932.
446. Minutes, 18 January 1961.
447. Minutes, 28 December 1931.
448. Minutes, 29 January 1932.
449. *Sydney Morning Herald,* 15 April 1930.
450. Minutes, 19 February 1932. Unfortunately, the loan agreement does not seem to have survived.
451. Minutes, 19 February 1932.
452. Peter Bell, *Robert Jowitt & Sons, the Limited Company.*
453. Minutes, 18 September 1931.
454. Minutes, 4 December 1931.
455. Minutes, 18 September 1931, 28 December 1931 & 29 January 1932.
456. Minutes, 30 September & 28 October 1932.
457. Minutes, 20 September 1935.
458. Minutes, 29 September 1932.
459. Minutes, 28 April 1933.
460. Minutes, 26 May 1933.
461. Peter Bell, *History of the Woolcombers' Mutual Association Limited,* pp. 19, 20.
462. Peter Bell, *History of the Woolcombers' Mutual Association Limited,* p. 25.
463. Peter Bell, *History of the Woolcombers' Mutual Association Limited,* p. 1.
464. Peter Bell, *History of the Woolcombers' Mutual Association Limited,* p. 14. See Appendix XII.
465. Minutes, 18 January 1935.
466. Peter Bell, *Robert Jowitt & Sons, the Limited Company.*
467. 1901 Census.
468. Minutes, 12 January 1940.
469. Minutes, 26 October 1939.
470. Minutes, 8 April 1932.
471. Minutes, 28 October 1938.
472. Minutes, 16 June 1939 & 15 September 1939.
473. Minutes, 17 July 194.
474. *Yorkshire Post,* 10 August 1940.
475. Minutes, 22 October 1937.
476. Minutes, 22 October & 26 November 1937.
477. Minutes, 21 January 1938.
478. Minutes, 4 March 1938.
479. Minutes, 8 April 1938.
480. Minutes, 27 May 1938.
481. Minutes, 24 June & 27 July 1938.
482. Minutes, 28 October 1938.
483. Minutes, 5 May 1939.
484. Minutes, 19 June 1939.
485. Minutes, 21 February 1951.
486. Exchange rate from *The Times,* 9 January 1951.

487. *The Times,* 1 October 1938.
488. Minutes, 9 October 1936.
489. Minutes, 28 May 1937.
490. Shay, 'Chamberlain's folly'.
491. Minutes, 8 April 1938.
492. Minutes, 16 September 1938.
493. Minutes, 5 May 1939.
494. Minutes, 26 October 1939.
495. Minutes, 24 September 1941.
496. Minutes, 14 October 1942.
497. Peter Bell, *Robert Jowitt & Sons, the Limited Company.*
498. Minutes, 12 January 1940.
499. Minutes, 13 August 1941, 24 September 1941 and 5 November 1941.
500. Minutes, 2 July 1941, 24 September 1941 & 5 November 1941.
501. Briggs, 'The Framework of the Wool Control'.
502. Minutes, 13 March, 1940
503. Minutes, 12 January 1940.
504. Minutes, 11 December 1940.
505. Minutes, 12 January 1940.
506. Minutes, 12 January 1940.
507. Murphy, 'Wartime Concentration of British industry'.
508. *The Times,* 14 March 1941.
509. Minutes, 13 August 1941.
510. Minutes, 16 October 1941.
511. Minutes, 5 November 1941.
512. Minutes, 19 December 1941.
513. Minutes, 5 Mmarch 1943.
514. Minutes, 2 June 1943.
515. Minutes, 6 December 1932.
516. Minutes, 12 January 1940.
517. Minutes, 6 June 1943.
518. Minutes, 13 March, 1940
519. Minutes, 5 June 1940.
520. Minutes, 21 August 1940.
521. Minutes, 22 January 1943.
522. *Barrier Miner* (Broken Hill, NSW), 11 March 1943.
523. Minutes, 2 June 1943.
524. Peter Bell, *Robert Jowitt & Sons, the Limited Company.*
525. Minutes, 5 June 1940.
526. Minutes, 9 April 1941.
527. Minutes, 5 March 1941.
528. Minutes, 9 March 1941.
529. Signed waiver included with minutes.
530. Minutes, 2 June 1943.
531. Minutes, 14 July 1943.

532. Minutes, 15 September 1943.
533. Minutes, 15 September 1943.
534. Minutes, 27 January 1944.
535. Sam Harland, Minutes, 31 May 1945.
536. Minutes, 13 September 1944.
537. Parliamentary debate reported in *The Times,* 20 November 1945.
538. It was left to others to accomplish the task which Hitler had not even contemplated. Any visitor to Leeds in the sixties or seventies, confronted by fields of bricks, the remnants of solidly-built terrace houses, would have found it hard to believe that his own countrymen rather than the Luftwaffe were responsible for this wanton destruction.
539. Minutes, 9 May 1946.
540. Minutes, 18 July 1946.
541. Minutes, 2 March 1949.
542. Minutes, 6 April 1949.
543. Minutes, 26 July 1949.
544. Minutes, 25 October 1950.
545. Minutes, 17 January 1951.
546. Minutes, 19 September 1951.
547. Minutes, 20 February 1952.
548. Minutes, 23 April 1952.
549. Minutes, 23 July 1952.
550. Minutes, 4 March 1953.
551. Minutes, 13 May 1953.
552. Minutes, 26 August 1953.
553. Minutes, 21 October 1953.
554. Minutes, 16 December 1953.
555. Minutes, 14 December 1954.
556. Minutes, 27 April 1955.
557. Minutes, 19 May 1955.
558. Probably influenza, although the Office of National Statistics figures seem to indicate that this was not exceptionally serious during the winter of 1955/6.
559. Minutes, 16 November 1955.
560. Minutes, 18 April 1956.
561. Minutes, 16 May 1956.
562. Minutes, 20 June 1956.
563. Minutes, 12 December 1956.
564. Minutes, 23 January 1957.
565. Minutes, 14 November 1956.
566. Minutes, 12 December 1956.
567. Minutes, 20 February 1957.
568. Minutes, 19 June 1957.
569. Minutes, 17 July 1957.
570. Minutes, 29 January 1958.
571. Minutes, 11 December 1957.

572. Minutes, 25 June 1958.
573. Minutes, 23 July 1958.
574. Minutes, 1 October 1958.
575. Minutes, 21 January 1959.
576. Minutes, 28 June & 26 July 1944.
577. Minutes, 29 September & 14 December 1948; Peter Bell, *History*.
578. Minutes, 1 March 1950.
579. Peter Bell, *History*.
580. Minutes, 22 June 1949.
581. Peter Bell, *History*.
582. Minutes, 2 March 1949.
583. Minutes, 6 April 1949.
584. Minutes, 26 July 1950.
585. Minutes, 13 May 1953.
586. Minutes, 7 June 1950, 27 September 1950, 25 October 1950, 6 December 1950, 17 January 1951, 21 February 1951, 14 March 1951, 17 April 1951 & 21 May 1951.
587. Minutes, 18 November 1959.
588. Minutes, 20 January 1960.
589. Minutes 5 April 1945
590. Minutes, 31 May 1945.
591. Minutes, 18 October 1945.
592. Minutes, 18 October 1945 & 23 January 1946.
593. Minutes, 18 July 1946.
594. Minutes, 23 November 1946.
595. Minutes, 17 December 1946.
596. Minutes, 5 November 1947.
597. Minutes, 18 December 1947.
598. Minutes, 29 January 1948.
599. Minutes, 29 January 1948.
600. Minutes, 13 July 1948.
601. Minutes, 3 November 1948.
602. Minutes, 26 October 1949.
603. Minutes, 14 December 1949.
604. Minutes, 6 December 1950 & 20 February 1952.
605. Minutes, 2 March 1949.
606. Minutes, 16 June 1954.
607. Minutes, 22 July 1959.
608. Approval was given for Mr A. A. Gibson to purchase a 10 horse-power Austin car for about £A750 – Minutes, 1 March 1950.
609. Minutes, 16 December 1953.
610. Minutes, 27 February 1946.
611. Barclays extended the overdraft to £800,000 in June and £850,000 in July – Minutes, 7 June & 20 July 1950.
612. Minutes, 6 December 1950.
613. Minutes, 1 March 1950.

614. Minutes, 29 March 1950.
615. Minutes, 1 March 1950.
616. Minutes, 7 June 1950.
617. The minutes for 16 December 1953 show the following share transfers of £1 cumulative preference shares were approved by the Board: from RBJ II and others, 10,000 to Commercial Union Assurance Co. Ltd, 5,000 to Guardian Assurance Co. Ltd, 2,400 to Atlas Electric & General Trust Ltd; from E. M. Jowitt, 10,000 to Prudential Assurance Co. Ltd, 5,000 to Industrial & General Trust Ltd, 2,168 to Trustees Corporation Ltd; from J. A. Jowitt, 2,000 to Atlas Electric & General Trust Ltd, 400 to Trustees Corporation Ltd; from Sam Harland, 10,000 to Industrial & General Finance Corporation, 600 to Atlas Electric & General Trust Ltd; from Richard McC. B. Jowitt, 5,000 to the Nineteen Twenty-Eight Investment Trust Ltd, 5,000 to Continental Union Trust Co. Ltd, 2,432 to the Trustees Corporation Ltd. The following new cumulative preference shares were issued at the same time: 15,000 to Industrial & Commercial Finance Corporation Ltd; 10,000 to Commercial Union Assurance Co. Ltd; 10,000 to Prudential Assurance Co. Ltd; 5,000 to Atlas Electric & General Trust Ltd; 5,000 to the Trustees Corporation Ltd; 5,000 to Guardian Assurance Co. Ltd; 5,000 to the Nineteen Twenty-Eight Investment Trust Ltd.
618. Minutes, 24 May 1950.
619. Minutes, 1 March 1950.
620. Minutes, 7 June 1950 & 27 September 1950.
621. Minutes, 6 December 1950.
622. Minutes, 24 January 1950.
623. Minutes, 18 January 1950.
624. Minutes, 3 March 1954.
625. Minutes, 20 February 1952.
626. Minutes, 21 January 1953.
627. Minutes, 29 October 1952.
628. Minutes, 13 December 1945.
629. Source: Reid family.
630. Minutes, 18 July 1946.
631. Minutes, 13 December 1945.
632. Minutes, 9 May 1946.
633. Minutes, 26 October 1949.
634. Minutes, 29 September 1953.
635. Minutes, 21 October 1953.
636. Minutes, 21 October 1953. Age calculated from 1901 census.
637. Minutes, 16 Decmber 1953.
638. Minutes, 25 November 1953.
639. Peter Bell, *Robert Jowitt & Sons, the Limited Company*.
640. Minutes, 21 October 1953.
641. Minutes, 25 November 1953.
642. Source: Reid family.

643. Minutes, 13 October 1954.
644. Minutes, 17 November 1954.
645. Minutes, 21 May 1958.
646. Minutes, 16 February 1955.
647. Minutes, 23 September 1959.
648. Minutes, 11 December 1957.
649. Minutes, 18 November & 16 December 1959.
650. Minutes, 30 May 1950.
651. Minutes, 6 September 1945.
652. Minutes, 20 February 1957.
653. Minutes, 18 July 1946 & 17 December 1946.
654. Minutes, 25 June 1952.
655. Minutes, 13 April 1947.
656. Minutes, 13 April 1947.
657. Minutes, 30 May 1950.
658. Minutes, 20 January 1954. Admittedly, the Directors were also questioning whether Meanwood was economic by December 1954 – Minutes, 14 December 1954.
659. Minutes, 5 June 1946.
660. Minutes, 20 November 1956.
661. Minutes, 11 December 1958.
662. Minutes, 21 January 1959.
663. Minutes, 18 November 1959.
664. Minutes, 20 January 1960.
665. Minutes, 14 September 1955 & 16 November 1955.
666. Minutes, 18 January 1956 & 14 March 1956.
667. Minutes, 20 June 1956.
668. Minutes, 23 April 1958.
669. Minutes, 27 May 1959.
670. Minutes, 16 December 1959.
671. Minutes, 27 May 1959.
672. Source: David Reid, personal communication.
673. Source: Tom Sykes.
674. Minutes, 19 June 1939.
675. Minutes, 4 May 1944.
676. Minutes, 18 December 1947.
677. Minutes, 18 December 1947.
678. Minutes, 3 November 1948 & 26 October 1949.
679. Minutes, 8 September 1954, 1 October 1958 & 16 January 1962.
680. Minutes, 20 January 1960.
681. Minutes, 27 April 1960.
682. Minutes, 18 May 1960.
683. Minutes, 21 June 1960.
684. Minutes 22 July 1953 & 27 July 1960.
685. Minutes, 27 July 1960.

686. Minutes, 20 January & 21 September 1960.
687. Minutes, 21 September 1960.
688. Minutes, 14 December 1960 & 18 January 1961.
689. Minutes, 14 December 1960.
690. Minutes, 14 December 1960.
691. Minutes, 18 January & 15 February 1961.
692. Minutes, 22 March 1961.
693. Minutes, 26 April 1961.
694. Minutes, 17 May 1961.
695. Minutes, 26 July 1961.
696. Minutes, 17 January 1962.
697. Minutes, 20 September 1961.
698. Minutes, 17 January 1962.
699. Minutes, 23 May 1962.
700. Minutes, 18 July 1962.
701. Minutes, 26 September 1962.
702. Minutes, 12 December 1962.
703. Minutes, 23 January 1963.
704. Minutes, 23 January 1963.
705. Minutes, 23 January & 24 April 1963.
706. Minutes, 22 May 1963.
707. Minutes, 24 July 1963.
708. Minutes, 11 December 1963.
709. Minutes, 23 October 1963.
710. Minutes, 19 November 1963.
711. Minutes, 11 December 1963.
712. Minutes, 11 December 1963.
713. Minutes, 29 January 1964.
714. Minutes, 22 April 1964.
715. Minutes, 3 June & 22 July 1964.
716. Minutes, 9 December 1964.
717. Minutes, 24 February 1965.
718. Minutes, 9 December 1964.
719. Minutes, 3 November 1964.
720. Minutes, 9 December 1964.
721. Minutes, 27 January 1965.
722. Minutes, 28 April 1965.
723. Peter Bell, *A History of the Woolcombers' Mutual Association Limited,* p. 68.
724. Minutes, 23/24 March 1965.
725. Minutes, 28 April 1965.
726. Minutes, 26 May 1965.
727. Minutes, 26 May 1965, my italics.
728. Minutes, 26 May 1965.
729. Minutes, 21 July 1965.
730. FTBJ personal reminiscence.

731. Source: F. T. B. Jowitt and P. J. M. Bell.
732. Minutes, 3 October 1965.
733. Minutes, 30 December 1965.
734. Minutes, 16 December 1959, 20 January 1960, 17 February 1960, 23 March 1960 & 18 May 1960.
735. Minutes, 14 December 1960.
736. Minutes, 14 December 1960.
737. The Perth and Fremantle offices having been merged in 1959.
738. Minutes, 27 September 1961.
739. Minutes, 1 November 1962.
740. Minutes, 13 December 1961.
741. Minutes, 26 April 1961.
742. Minutes, 17 January 1962.
743. Minutes, 21 February 1962.
744. Minutes, 21 March 1962.
745. Minutes, 24 October 1962.
746. Minutes, 23 January 1963.
747. Minutes, 20 February, 24 April & 22 May 1963.
748. Minutes, 3 November 1964.
749. Minutes, 23 March 1965.
750. Minutes, 13 April 1966.
751. Minutes, 23 March 1965.
752. Minutes, 28 April 1965.
753. Minutes, 21 July 1965.
754. Minutes, 17 December 1965.
755. Minutes, 13 February 1969.
756. Minutes, 21 July 1965.
757. Minutes, 6 October 1965.
758. Minutes, 22 June 1966.
759. Minutes, 26 January 1966.
760. Minutes, 22 June 1966.
761. Minutes, 27 May 1959 & 14 June 1959.
762. Minutes, 22 March 1961.
763. Minutes, 22 March 1961.
764. Minutes, 26 April 1961.
765. Minutes, 26 July 1961.
766. Minutes, 21 August 1961.
767. Minutes, 21 August 1961.
768. Minutes, 1 November 1961 & 13 December 1961.
769. Minutes, 23 May 1962.
770. Minutes, 23 May 1962.
771. Minutes, 18 July 1962.
772. Minutes, 26 September 1962 & 24 October 1962.
773. Minutes, 1 March 1963.
774. Source: Peter Bell.

775. Minutes, 20 February 1963 & 24 April 1963.
776. Minutes, 20 February 1963 & 24 April 1963.
777. Minutes, 25 September 1963.
778. Minutes, 25 September 1963 & 26 February 1964.
779. Minutes, 3 June 1964.
780. Minutes, 3 June 1964.
781. Minutes, 22 July 1964.
782. Minutes, 30 September 1964.
783. Minutes, 27 January 1965.
784. Minutes, 24 February 1965.
785. Minutes, 6 October 1965.
786. Minutes, 30 December 1965.
787. Minutes, 18 May 1960.
788. Minutes, 26 July 1961.
789. Minutes, 27 September 1961.
790. Minutes, 1 November 1961.
791. Minutes, 26 September 1962.
792. Minutes, 23 October 1963.
793. Minutes, 20 March 1963.
794. Minutes, 26 January 1966.
795. Minutes, 26 January 1966.
796. Minutes, 22 June 1966.
797. Minutes, 27 January 1967.
798. Longstanding restrictive agreements, which one likes to think would be illegal now, had divided up the trade in such a way that some fellmongeries, including Meanwood, were only allowed to process imported skins.
799. Minutes, 28 June & 27 September 1967.
800. Source: F. T. B. Jowitt.
801. Minutes, 14 December 1966.
802. Minutes, 16 September 1966, 14 December 1966, 22 March 1967 & 28 June 1967.
803. Minutes, 27 January 1967.
804. Minutes, 28 June & 27 September 1967, 5 April 1968.
805. Minutes, 31 January 1968.
806. Minutes, 18 December 1968 & 25 June 1969.
807. Minutes, 18 December 1968, 19 March 1969.
808. Minutes, 25 June & 29 October 1969.
809. Minutes, 17 December 1969.
810. Source: Peter Bell.
811. Minutes, 18 December 1968.
812. Minutes, 22 June 1966.
813. Minutes, 27 January 1967.
814. Minutes, 6 March 1968.
815. Minutes, 23 October 1968.
816. Minutes, 8 July 1970.

817. Source: F. T. B. Jowitt.
818. Minutes, 27 January 1967.
819. Minutes, 31 January 1968.
820. Minutes, 13 February 1969.
821. Minutes, 22 April 1968.
822. Minutes, 28 June & 23 October 1968.
823. Minutes, 19 March 1969.
824. Minutes, 29 October 1969.
825. Minutes, 22 June 1966.
826. Minutes, 25 June 1969.
827. Minutes, 30 December 1965.
828. Minutes, 7 February, 19 March & 25 June 1969.
829. Minutes, 6 October 1965.
830. Minutes, 8 April 1970.
831. Minutes, 8 April 1970.
832. Minutes, 8 June 1970.
833. Minutes, 8 July 1970.
834. Minutes, 8 July & 18 December 1970.
835. Minutes, 15 December 1971.
836. Minutes, 17 September 1974.
837. Minutes, 23 June 1976.
838. Minutes, 16 December 1970 & 17 March 1971.
839. Minutes, 15 December 1971.
840. *London Gazette,* 29 March 1973.
841. Minutes, 17 March 1971.
842. Minutes, 15 December 1971
843. Minutes, 20 February 1974.
844. Minutes, 12 May 1978.
845. Minutes, 22 March & 19 April 1972.
846. Minutes, 21 June 1972.
847. Minutes, 1 November 1973 & 18 December 1974.
848. Minutes, 11 December 1975, 31 March 1976 & 17 December 1976.
849. Minutes, 26 September 1977, 12 May 1978 & 15 December 1978.
850. Minutes, 14 December 1979.
851. Minutes, 17 September 1982.
852. Minutes, 28 January 1983.
853. Minutes, 18 April 1986.
854. Minutes, 30 October & 11 December 1987.
855. Minutes, 21 June 1972, 21 March 1973.
856. Bank of England statistics – http://www.bankofengland.co.uk/.
857. Minutes, 26 June 1974.
858. Minutes, 11 December 1972.
859. Minutes, 26 January 1973.
860. Minutes, 26 June 1974.
861. Minutes, 18 December 1974.

862. Minutes, 19 March 1975.
863. Minutes, 12 May & 22 September 1978.
864. Minutes, 18 May & 28 September 1979.
865. Minutes, 14 May 1980.
866. Minutes, 3 February 1983.
867. Minutes, 23 January 1983.
868. Minutes, 22 June, 21 September & 30 October 1984.
869. Minutes, 19 March & 26 June 1975.
870. Minutes, 17 September 1975.
871. Minutes, 31 March 1976.
872. Minutes, 16 February 1977.
873. Minutes, 6 May 1977.
874. Minutes, 28 September 1979.
875. Minutes, 14 December 1979.
876. Minutes, 28 March 1980.
877. Minutes, 6 June 1980.
878. Minutes, 16 January 1981.
879. Minutes, 12 September 1980.
880. Minutes, 16 January 1981.
881. Minutes, 10 July 1981.
882. Minutes, 3 February 1983.
883. Minutes, 28 January 1983.
884. Minutes, 10 June 1983.
885. Minutes, 15 September 1983.
886. Notice of creditors' meeting to be held on 5 October – *London Gazette,* 5 September 1984.
887. Minutes 27 February, 31 March & 15 December 1976.
888. Minutes, 15 September 1976.
889. Minutes, 15 December 1977.
890. Minutes, 31 March 1976.
891. Minutes, 23 June 1976.
892. Minutes, 15 December 1977.
893. Minutes, 28 September & 14 December 1979.
894. Minutes, 28 March 1980, 6 June 1980, 12 September 1980 & 16 January 1981.
895. Minutes, 17 September 1982.
896. Minutes, 10 June 1983.
897. Minutes, 27 January 1984.
898. Minutes, 19 September 1986.
899. Minutes, 12 December 1986.
900. Minutes, 11 December 1987, 18 March 1988 & 23 September 1988.
901. Minutes, 12 May 1989.
902. Minutes, 16 December 1988.
903. Minutes, 29 September 1989.
904. Minutes, 19 October 1990, 28 January 1991 & 22 July 1991; sale agreement, 31 May 1991.

905. A company set up by F. T. B. Jowitt's wife's cousins, Michael ('Harpo') and Ian ('the Bloater') Brackenbury.
906. Minutes, 16 January 1981.
907. Minutes, 6 June 1980.
908. Minutes, 24 April 1981.
909. Minutes, 10 July 1981.
910. Minutes, 27 January 1984.
911. Minutes, 22 June 1984.
912. Minutes, 21 September 1984.
913. Minutes, 18 April 1986.
914. Minutes, 19 September 1986.
915. Minutes, 22 July 1959.
916. Minutes, 28 June 1967.
917. Minutes, 15 December 1978.
918. Minutes, 19 October 1990 & 28 January 1991.
919. Activity on the Sydney Futures Market was handled from Bradford.
920. Minutes, 24 April 1981.
921. Minutes, 10 July 1981.
922. Minutes, 22 June 1984, 21 September 1984 & 10 May 1985.
923. Minutes 19 September 1986.
924. The minutes say 'Bank of England' but it seems clear from the context that the South African Reserve Bank is meant, as the Bank of England would have no interest in blocking money transfers from South Africa to the United Kingdom.
925. Minutes,12 May & 15 December 1978.
926. Minutes, 24 April 1981.
927. Minutes, 28 January, 7 March & 10 June 1983.
928. Minutes, 8 July 1970.
929. Minutes, 17 March 1971.
930. Minutes, 12 May 1971.
931. Minutes, 12 May 1971.
932. Minutes, 14 July 1971.
933. Minutes 15 December & 12 May 1971.
934. Minutes, 15 December 1971.
935. Minutes, 18 May 1979.
936. Minutes, 22 September 1978.
937. Minutes, 28 March 1980.
938. Minutes, 12 September 1980.
939. Minutes, 28 January 1991.
940. Minutes, 11 December 1972.
941. Minutes, 15 December 1971.
942. *The Times,* 29 March 1972.
943. Minutes, 21 June 1972.
944. Minutes, 11 December 1972.
945. Minutes, 27 June 1973.

946. Minutes, 26 June 1974.
947. Minutes, 18 December 1974 & 17 September 1975.
948. Minutes, 11 December 1975.
949. Minutes, 17 December 1976.
950. Minutes, 6 May 1977.
951. Minutes, 12 May 1978.
952. Minutes, 27 September 1978.
953. Minutes, 15 December 1978.
954. Minutes, 10 July 1981.
955. Minutes, 18 January, 10 May & 20 September 1985.
956. Minutes, 10 January 1986.
957. Minutes, 18 September 1987.
958. Minutes, 11 December 1987.
959. Minutes, 23 September & 16 December 1988
960. Minutes, 22 October 1993. Remarkably, though, the mill manager. Mr S. H. Lambert, appealed against the findings of Her Majesty's Inspectorate of Pollution and was congratulated by that body for the quality of his application – Minutes, 22 April 1994.
961. Minutes, 18 October 1991.
962. Minutes, 15 December 1989, 18 September 1992.
963. Minutes, 2 April 1993.
964. Minutes, 21 April 1995.
965. Minutes, 26 October 1995.
966. Minutes, 11 January 1996.
967. Minutes, 19 April 1996.
968. Minutes, 14 March 1997.
969. Minutes, 5 September 1997.
970. Minutes, 17 April 1998.
971. Minutes, 4 December 1998.
972. Minutes, 4 December 1998.
973. Sources: Peter Bell; minutes, 27 April 2000.
974. Quoted in advice given by David Casement QC enclosed in a letter from the solicitors Lee & Priestly to Mr F. T. B. Jowitt, 2 July 2008.
975. Advice of David Casement QC enclosed in a letter from Lee and Priestly, solicitors, to Mr F. T. B. Jowitt, 2 July 2008. This advice relates to a later case which grew out of the original one.
976. The only instance of one of the family failing to pay debts which I have been able to find is the case of Richard Jowitt I whose estate on his death in 1696 was insufficient to pay his creditors in full.
977. The Board then consisted of: F. T. B. Jowitt (Chairman); his brother W. J. B. Jowitt; his son W. R. B. Jowitt; R. B. Jowitt II's son-in-law R. J. Heaton; and N. Matthews, the Company Secretary.
978. *Leeds Mercury,* 18 May 1850.
979. From William Cowper, *The Progress of Error.*
980. John Jowitt jr, Diary for 1835–6.

981. Robert Arthington senior is said to have given up the brewing business on conscientious grounds – *Leeds Mercury,* 11 October 1900. Indeed, according to Brian Stanley, 'The legacy of Robert Arthington,' he adopted strict temperance principles.

982. At 2 Hunslet Road (by 1881 it had been renumbered 3) – Census 1871, 1881. The renumbering is evident from the fact that his neighbour, Grace Wightman, is also one number higher in the 1881 census.

983. *New York Times,* 30 December 1900.

984. *Leeds Mercury,* 4 December 1900.

985. *New York Times,* 30 December 1900.

986. Unless he moved there after the 1881 census, this is clearly wrong. The newspaper reports seem generally rather confused, probably because of Arthington's reclusive nature. The *Leeds Mercury,* which also refers to him living in Headingly, calls Phoebe (Phebe on the census) his mother. She was actually his sister.

987. *North-Eastern Daily Gazette,* 11 October 1900.

988. *Leeds Mercury,* 11 October 1900.

989. Brian Stanley, 'The Legacy of Robert Arthington'.

990. *Leeds Mercury,* 17 October 1900.

991. *Daily News,* 18 December 1900.

992. Brian Stanley, 'The Legacy of Robert Arthington'.

993. Gray, *A History of Southern Sudan,* p. 168.

994. Pachuau, 'Church-mission dynamics in northeast India'.

995. *Morning Post,* 14 June 1858.

996. *Daily News,* 21 December 1863.

997. *Daily News,* 29 August 1966.

998. Service record.

999. Source: the Reverend David Jowitt.

1000. Obituary, *The Times,* 17 August 1957.

Bibliography

Anning, Stephen Towers, *The General Infirmary At Leeds: The Second Hundred Years.* Edinburgh & London: E. & S. Livingstone, 1963.

Anon, *An Account of the Constitution and Present State of Great Britain, Together With a View of Its Trade, Policy, and Interest, Respecting Other Nations, & of the Principal Curiosities of Great Britain and Ireland,* London: printed for J. Newbery, 1759.

———, *A History of the Town and Parish of Leeds, Compiled From Various Authors. To Which is Added, a History of Kirkstall Abbey,* Leeds: sold by all booksellers, 1797.

———, *Leeds and Bradford, Dewsbury, Batley, Keighley, &c.: England's Great Manufacturing Centres. Vol. Yorkshire; pt.1,* London: Historical Publishing Company, 1888.

———, *Oxford Dictionary of National Biography* [Online Resource], Oxford: Oxford University Press, 2004.

Barnard, A, 'Wool Buying in the Nineteenth Century: A Case Study,' *Bulletin of Economic Research* 8 (1), 1956, pp. 1–12.

Bates, Elisha. *An Examination of Certain Proceedings and Principles of the Society of Friends, etc,* St. Clairsville: For the Author, 1837.

Birchall, John Dearman & Emily, *The Diary of a Victorian Squire: Extracts From the Diaries and Letters of Dearman & Emily Birchall,* ed. David Verey, Gloucester: Alan Sutton, 1983.

Bell, Peter J. M., *Robt. Jowitt & Sons, the Limited Company 1919 to 1957,* Bradford, [n.d.].

———, *A History of the Woolcombers' Mutual Association Limited 1933 to 1994,* Cross Hills: Peter J. M. Bell, 2000.

Benson, George F., *Historical Record of the Edwardsburg and Canada Starch Companies,* Montreal: Canada Starch, 1958.

Benson, R. Seymour, *Photographic Pedigree of the Descendants of Isaac & Rachel Wlson 1740,* Middlesbrough: William Appleyard & Sons, 1912.

Beresford, Maurice Warwick, *Leeds Chamber of Commerce,* Leeds: Leeds Incorporated Chamber of Commerce, 1951.

Besse, Joseph, *A Collection of the Sufferings of the People Called Quakers,* London: Luke Hinde, 1753.

Billam, Francis T., *A Walk Through Leeds, Or Stranger's Guide to Every Thing Worth Notice in That Ancient and Populous Town; With an Account of the Woollen Manufacture of the West-Riding of Yorkshire,* Leeds: printed by J. H. Leach, 1806.

———. *Walks Through Leeds; Or the Stranger's Companion to the Public*

Buildings, Churches, Chapels, Charitable Institutions, &c. In That Ancient and Populous Town; and Various Historical Occurences Connected Therewith, Leeds: printed and sold by John Heaton, 1835.

Birchall, Emily, *Wedding Tour: January–June 1873 and Visit to the Vienna Exhibition,* ed. David Verey, Gloucester: Alan Sutton, 1985.

Birchall, Samuel, A *Descriptive List of the Provincial Copper Coins Or Tokens, Issued Between the Years 1786 and 1796, Arranged Alphabetically,* Leeds: Printed for S. Birchall and Sold by Henry Young, Ludgate Hill, London, 1796.

Bischoff, James, *A Comprehensive History of the Woolen and Worsted Manufacturers, and the Natural and Commercial History of Sheep: From the Earliest Records to the Present Time,* London: By Smith, Elder and Co., and Baines and Newsome, 1842.

Brigg, A., 'The Framework of the Wool Control.' *Oxford Economic Papers* 8, 1947, pp. 18–45.

Chapman, Stanley D., 'Fixed Capital Formation in the British Cotton Industry, 1770–1815.' *Economic History Review* 23 (2), 1970, pp. 235–66.

Coleridge, Samuel Taylor, *Collected Letters of Samuel Taylor Coleridge,* ed. Earl Leslie Griggs, Vol. 3, Oxford: Clarendon Press, 1959.

Cooley, Arnold James, *A Cyclopaedia of Several Thousand Practical Receipts, and Collateral Information in the Arts, Manufactures, and Trades, Including Medicine, Pharmacy, and Domestic Economy,* New York: D. Appleton & Company, 1846.

Costello, John, *Mask of Treachery: Spies, Lies and Betrayal,* New York: Warner Books, 1989.

Crewdson, Isaac, *A Beacon to the Society of Friends,* London: Hamilton, Adams & Co., 1835.

Cudworth, William, *Worstedopolis: A Sketch History of the Town and Trade of Bradford, the Metropolis of the Worsted Industry Compiled Expressly for the Bradford Manufacturing Company* (Reprint of 1888 Edn), Bradford: Old Bradfordian Press, 1997.

Defoe, Daniel, *A Tour Thro' the Whole Island of Britain, Divided Into Circuits Or Journeys,* London: G. Strahan, 1724.

Derek Fraser, ed., *A History of Modern Leeds,* Manchester: Manchester University Press, c1980., 1980.

Forster, William, *Memoirs of William Forster,* ed. Benjamin Seebohm, London: Alfred W. Bennett, 1865.

Foster, Sandys B., *The Pedigrees of Jowitt, Formerly of Churwell, Yorks, and Now of Harehills, Leeds, and the Families Connected with Them,* [n.p.]: printed for private circulation, 1890.

Gerhold, Dorian, 'Productivity Change Before and After Turnpiking, 1690–

1840,' *Economic History Review,* 49 (3), 1996.

Gillman, James, *The Life of Samuel Taylor Coleridge,* London: W. Pickering, 1838.

Gray, Richard, *A History of Southern Sudan 1839–1889,* London: Oxford University Press, 1961.

Hergeth, Helmut H. A., 'Compact Or Fancy,' *Textile World,* 152 (8), 2002, pp. 40–43.

Hirst, William, *History of the Woollen Trade for the Last Sixty Years; Shewing the Advantages Which the West of England Manufacturers Had Over Those of Yorkshire Up to 1813, How These Were Gradually Overcome, Until 1818, When a Challenge Was Received and Accepted for the Author in London. To Place His Goods in Competition With Those of the West of England; Commencing With a Memoir of the Author, etc.,* Leeds: by S. Moody, 1844.

Hopper, Andrew, 'The Farnley Wood Plot and the Memory of the Civil Wars in Yorkshire,' *The Historical Journal,* 45 (2), 2002, pp. 281–303.

Commons, House of, *Minutes of Evidence Taken before the Select Committee Appointed to Enquire into the Present State of Manufactures, Commerce and Shipping, 1833,* London: Printed by order of the House of Commons, 1833.

Housman, John, *A Descriptive Tour, and Guide to the Lakes, Caves, Mountains, and Other Natural Curiosities, in Cumberland, Westmoreland, Lancashire, and a Part of the West Riding of Yorkshire,* Carlisle: Printed by F. Jollie and sold by C. Law, London, 1800.

Hudson, Pat, *The West Riding Wool Textile Industry: A Catalogue of Business Records From the Sixteenth to the Twentieth Century,* Pasold Occasional Papers, 3, Edington: Pasold Research Fund, 1975.

———, The Genesis of Industrial Capital: A Study of the West Riding Wool Textile Industry, C. 1750–1850. Cambridge and New York: Cambridge University Press, 1986.

———, The Limits of Wools. Cardiff Historical Papers, Cardiff: Cardiff University, 2007.

Huurdeman, Anton A., *The Worldwide History of Telecommunications,* Hoboken, NJ: John Wiley & Sons, 2003.

Jowitt, John, Robert Benson Jowitt et al., *Reminiscences of John Jowitt, by His Children,* Gloucester: John Bellows, 1889.

Jowitt, Robert, *Thoughts on Water-Baptism,* London: Darton and Harvey, 1837.

Keynes, John Maynard, *The Economic Consequences of the Peace,* London: Macmillan, 1919.

———, Essays in Persuasion, New York: Harcourt, Brace and Company, 1932.

Kopf, M. E., 'On an Improved Process of Manufacturing Soda and Sulphuric

Acid,' *The Chemist,* 4, 1857, pp. 33–37.

Leonard, Raymond W,. *Secret Soldiers of the Revolution: Soviet Military Intelligence,* 1918–1933. Westport CT: Greenwood Press, 1999.

Mercer, Thomas, *A Vocabulary, Latin and English, on a New Plan Being a Selection of Words From the Purest Roman Authors, Progressively Arranged According to the Frequency of Their Occurrence in a Regular Course of Classical Tuition, and Distinguished by Their Syllabic Quantities,* London: Printed for Longman, Hurst, Rees, Orme, and Browne; and Isaac Nichols, Moot Hall, Leeds, 1817.

Morris, R. J., *Men, Women and Property in England, 1780–1870: A Social and Economic History of Family Strategies Amongst the Leeds Middle Classes,* Cambridge: Cambridge University Press, 2005.

Mortimer, Jean E., and Russell, *Leeds Friends' Minute Book, 1692–1712,* [Leeds]: Yorkshire Archaeological Society, 1980.

Murphy, Mary E., 'Wartime Concentration of British Industry,' *Quarterly Journal of Economics,* 57 (1), 1942, pp. 129–41.

Nicholson, Cornelius, *The Annals of Kendal: Being a Historical and Descriptive Account of Kendal and Its Environs: With Biographical Sketches of Many Eminent Personages Connected With the Town,* 2nd edn, London: Whitaker & Co., 1861.

O'Donoghue, Jim, Louise Goulding, and Grahame Allen, 'Consumer Price Inflation Since 1750,' *Economic Trends,* 604, 2004, pp. 38–46.

Odlyzko, A. M., 'This Time is Different: An Example of a Giant, Wildly Speculative, and Successful Investment Mania,' *BE Journal of Economic Analysis and Policy,* 10 (1), 2010.

Pachuau, Lalsangkima, 'Church-Mission Dynamics in Northeast India,' *International Bulletin of Missionary Research,* 27 (4), 2003.

Palmer, Joseph, *A Fortnight's Ramble to the Lakes in Westmoreland, Lancashire, and Cumberland, By a Rambler,* London: printed for Hookham and Carpenter, New and Old Bond Street, 1792.

Poulter, J. D., *An Early History of Electricity Supply: The Story of the Electric Light in Victorian Leeds,* IEE history of technology series, London: Peregrinus on behalf of the Institution of Electrical Engineers, 1986.

Radcliffe, Ann Ward, *A Journey Made in the Summer of 1794, through Holland and the Western Frontier of Germany, with a Return Down the Rhine: To Which Are Added Observations During a Tour to the Lakes of Lancashire, Westmoreland, and Cumberland,* London: G. G. and J. Robinson, 1795.

Roberts, R. O., 'Bank of England Branch Discounting, 1826–59,' *Economica,* new Series 25 (99), 1958, pp. 230–45.

Ryley, John, *The Leeds Guide: Including a Sketch of the Environs and Kirkstall Abbey,* Leeds: Printed By Edward Baines for the Author; and Sold By R.

Brown, 1806.

Salzedo, Simon, and Peter Brunner, *Briefcase on Contract Law,* London: Cavendish Publishing, 1999.

Shay, Robert P. 'Chamberlain's Folly: The National Defence Contribution of 1937,' Albion: a Quarterly Journal Concerned with British Studies, 7 (4), 1975, pp. 317–27.

Sheffield, John, *The Works of John Sheffield, Earl of Mulgrave, Marquis of Normanby, and Duke of Buckingham,* London: Printed for T. Wotton, 1740.

Silberling, Norman J., 'British Prices and Business Cycles, 1779–1850,' *Review of Economic Statistics,* 5, Supp. 2, 1923, pp. 223–47.

Friends, Society of, *Extracts from the Minutes and Proceedings of the Yearly Meeting of Friends, Held in London; together with Testimonies, Concerning Deceased Friends Presented to it,* London: Edward Marsh, 1863

——, Friends' Historical Society, *Journal of the Friends' Historical Society, London,* various dates.

——, *The Annual Monitor,* [Obituary of the Members of the Society of Friends in Great Britain and Ireland], York, various dates.

——, *The Inquirer,* various dates.

Somervell, John, *Isaac and Rachel Wilson, Quakers, of Kendal, 1714-85,* London: Swarthmore Press, 1924.

Stanley, Brian, 'The Legacy of Robert Arthington,' *International Journal of Missionary Research,* 22 (4), 1998.

Taylor, John, *The Praise of Hemp-Seed With the Voyage of Mr. Roger Bird and the Writer Hereof, in a Boat of Brown-Paper, From London to Quinborough in Kent. As Also, a Farewell to the Matchlesse Deceased Mr. Thomas Coriat. Concluding With the Commendations of the Famous Riuer of Thames,* London: Printed. for H. Gosson, 1620.

Teale, Thomas P., *A Historical Sketch of the General Infirmary at Leeds, 1767–1916,* Leeds: Jackson, 1917.

Thom, David, *Sketch of the Life and Character of the Late Samuel Mcculloch, Esq., Liverpool: Member of the Royal College of Surgeons,* Liverpool: D. Marples, 1853.

Thoresby, Ralph, *Ducatus Leodiensis: Or, the Topography of the Ancient and Populous Town and Parish of Leedes, and Parts Adjacent in the West Riding of the County of York, With the Pedigrees of Many of the Nobility and Gentry, and Other Matters Relating to Those Parts, Extracted From Records, Original Evidences, and Manuscripts, to Which is Added a Catalogue of His Museum, With the Curiosities Natural and Artificial, and the Antiquities, Particularly the Roman, British, Saxon, Danish, Norman, and Scotch Coins, With Modern Medals, Also a Catalogue of Manuscripts, the Various Editions of the Bible, and of Books Published in the Infancy of the Art of Printing, With an Account of Some Unusual Accidents That Have Attended Some Persons, Attempted After the Method of Dr. Plot,* London: Printed for Maurice Atkins, and sold by Edward Nutt, 1715.

——, *The diary of Ralph Thoresby, F.R.S., Author of the Topography of Leeds. (1677-1724.) Now first published from the original manuscript,* ed. Rev. Joseph Hunter, Vol. 1, London: Henry Colburn and Richard Bentley, 1830.

Tomkins, John, *Piety Promoted, in a Collection of Dying Sayings of Many of the People Called the Quakers: With a Brief Account of Some of Their Labours in the Gospel, and Sufferings for the Same. A New and Complete Edition, Comprising the Eleven Parts* [By J. Tomkins, J. Field, J. Bell, T. Wagstaffe, J. G. Bevan, J. Forster] *Heretofore Separately Published,* ed. William & Thomas Evans, Philadelphia: [William & Thomas Evans], 1854.

Tregoning, David, and Hugh Cockerell, *Friends for Life: Friends' Provident Life Office 1832–1982,* London: H. Melland, 1982.

Manchester, Victoria University of, *The Victoria University of Manchester: Register of Graduates up to July 1st,* 3rd edn, 1908, Manchester: Sherratt & Hughes, 1908.

Wails, Isabel, *A Warning to the Inhabitants of Leeds and All Others in Cities, Towns and Villages, Who Have Willfully Been Persecuting the People of the Lord: Whom He Hath Called By His Eternal Spirit to Magnifie Himself in, and to Testifie for Truth and Righteousness, and Against All Ungodly Works, and Workers Thereof,* [London]: [n.p.], 1685.

Wake, Jehanne, *Kleinwort, Benson: A History of Two Families in Banking,* Oxford: Oxford University Press, 1997.

Ward, J. T., *The Factory Movement 1830–1855,* London: Macmillan, 1962.

Weddle, Meredith Baldwin, *Walking in the Way of Peace: Quaker Pacifism in the Seventeenth Century,* New York: Oxford University Press, 2001.

Wilbur, John, *Journal of the Life of John Wilbur, a Minister of the Gospel in the Society of Friends: With Selections From His Correspondence, &c.,* Providence, Rhode Island: G. H. Whitney, 1859.

Wordsworth, William, *Memorials of a Tour on the Continent, 1820,* London: Longman, Hurst, Rees, Orme, and Brown, 1822.

Young, Arthur, *Observations on the Present State of the Waste Lands of Great Britain. Published on the Establishment of a New Colony on the Ohio,* London: W. Nicoll, 1773.

315

Index

Reversed out page numbers (e.g.) indicate illustrations, underlined page numbers (e.g. <u>32</u>) indicate extracts of family trees.

www.ingramcontent.com/pod-product-compliance
Lightning Source LLC
Chambersburg PA
CBHW070944150426
42812CB00066B/3286/J